SUSTAINABLE POWER COOLING METHODS FOR TROPICAL CLIMATES & PLACES

ISHAN

Contents

CONTENTS I

ACKNOWLEDGEMENTS V

LIST OF FIGURES.. VI

LIST OF TABLES.. IX

ACRONYMS X

LIST OF SYMBOLS XI

ABSTRACT XIII

CHAPTER 1   INTRODUCTION... 1

1.1   OUTLINE OF THE THESIS 2

CHAPTER 2   LITERATURE REVIEW 5

2.1   THERMAL COMFORT MODELS 5

2.1.1 Indian thermal comfort standards 7

2.2   PASSIVE COOLING 8

2.2.1 Natural ventilation 9

2.2.2 Controlled night time cooling....................................12

2.2.3 Earth Air Heat exchanger (EAHE) 13

2.3   INDOOR TEMPERATURE MODEL...................................15

2.3.1 Analytical building heat transfer models......................15

2.4   ARTIFICIAL NEURAL NETWORKS (ANN)..22

2.4.1 Summary 27

CHAPTER 3   OBJECTIVES OF THE THESIS **28**

CHAPTER 4   THERMAL COMFORT MODELS **29**

4.1   PHYSIOLOGICAL BASIS OF THERMAL COMFORT...29

4.1.1 Environmental thermal comfort factors..31

4.2   THERMAL COMFORT MODELS...33

4.2.1 Victor Olgyay Bioclimatic chart...34

4.2.2 Fanger heat balance (comfort) equation..34

4.3   ADAPTIVE THERMAL COMFORT MODEL..36

4.4   IDENTIFICATION OF SUITABLE ADAPTIVE THERMAL COMFORT MODEL.......................39

CHAPTER 5   PASSIVE AND NATURAL COOLING..**41**

5.1   HEAT GAIN REDUCTION AND MODULATION...42

5.1.1 Solar Radiation on the Building Envelope...42

5.1.2 Building Length (L) / Width (W)..44

5.1.3 Building Orientation 45

5.1.4 Building Materials..46

5.2   EARTH AIR HEAT EXCHANGER (EAHE)...48

5.2.1 EAHE modelling...50

5.2.2 Experiments with EAHE..51

CHAPTER 6   NATURAL VENTILATION 55

    6.1   WIND DRIVEN NATURAL VENTILATION 56

    6.2   WORK CARRIED OUT ON WING WALLS (TVK SUSHIL KUMAR) 58

      6.2.1 Validation 59

      6.2.2 Effect of Wing Walls on Natural Ventilation 60

      6.2.3 Study of Air Flow Pattern for Different Wing Wall Positions 61

      6.2.4 Effect of Wing Length on Average Inlet Velocities 63

      6.2.5 Wing Wall Height 63

      6.2.6 Inverted 'L' Type Wing Wall 64

CHAPTER 7   NOCTURNAL COOLING 66

      7.1.1 Air Flow 66

      7.1.2 Temperature differential and Thermal mass 67

      7.1.3 Convective Heat Transfer Coefficient (CHTC) 68

    7.2   BARRIERS TO NIGHT VENTILATION 68

    7.3   WORK CARRIED OUT ON ASCERTAINING THE NOCTURNAL VENTILATION FOR

      COIMBATORE. 69

CHAPTER 8   CONTROLLED NOCTURNAL COOLING 71

CHAPTER 9   PROPOSED ANALYTICAL AND ANN BASED BUILDING

      HEAT TRANSFER MODEL 75

    9.1   SOL-AIR TEMPERATURE 76

    9.2   THE PROPOSED MODEL 76

9.3   VALIDATION OF THE PROPOSED ANALYTICAL MODEL...79

9.4   ARTIFICIAL NEURAL NETWORK (ANN) BASED INDOOR TEMPERATURE MODEL...........80

   9.4.1 Validation of the ANN based model..81

CHAPTER 10   CONTROL LAW TO MAINTAIN INDOOR TEMPERATURES WITHIN ADAPTABLE COMFORT LIMITS...............................82

10.1   DEVELOPMENT OF THE CONTROL LAW 83

CHAPTER 11   RESULTS AND CONCLUSIONS..95

11.1   THERMAL COMFORT MODEL..95

11.2   IMPROVEMENTS IN PASSIVE AND NATURAL COOLING METHODS 96

   11.2.1 Optimization of building envelope...96

   11.2.2 Earth Air Heat Exchanger (EAHE)...96

   11.2.3 Improvement of natural ventilation with wing walls 96

   11.2.4 Study on night time ventilation (nocturnal cooling) 97

11.3   INTEGRATION OF DIFFERENT PASSIVE COOLING TECHNIQUES........................97

   11.3.1 Formulation of analytical and ANN based model...97

   11.3.2 Development of control law..98

   11.3.3 Limitation and scope for future work...98

REFERENCES 99

LIST OF PUBLICATIONS..107

Acknowledgements

I am indebted to the Almighty, my late grandparents and parents for their blessings in carrying out this research work.

In the year 2011, our Chancellor Mata Amritanandamayi, gave a clarion call to focus research on energy saving methods, for the benefit of the society. This work is an attempt to fulfil her vison.

I gratefully acknowledge Amrita Vishwa Vidyapeetham, Coimbatore, for providing the right atmosphere and resources toward my research.

A heartful thanks to Dr. J.Chandrasekar, my guide who gave me critical remarks, valuable suggestions and freedom to choose my area of research, which opened up new vistas for me.

I also take this opportunity to extend my sincere gratitude towards my thesis committee members: Dr. V. Sivadas, Dr. Balajee. R. and Dr. K.K. Sasi for their insightful thoughts and valuable suggestions.

I would like to express my heart felt gratitude to my wife Sumathi S., for the immense support and for being there by my side at all times.

Thanks to my daughter Ar. Hansika S. for building my house "Villa-de-Sudiksha", based on the work carried out in this Thesis, which will **allow me to live my Thesis for the rest of my life.**

Last but not the least, my son Manas S. for helping me in developing codes, fabricating the system and typing this report, in spite of his busy academic commitments.

List of Figures

FIGURE 1.1 FLOW CHART OF THESIS 4

FIGURE 2.1: DIURNAL VARIATION OF INDOOR AND OUTDOOR TEMPERATURES 18

FIGURE 2.2: VARIATION OF TIME LAG AS A FUNCTION OF THICKNESS AND DENSITY 19

FIGURE 2.3: VARIATION OF DECREMENT FACTOR AS A FUNCTION OF THICKNESS AND DENSITY 19

FIGURE 2.4: VARIATION OF THERMAL RESISTANCE WITH WIND SPEED 22

FIGURE 2.5: PERCEPTRON WITH THREE INPUTS AND ONE OUTPUT 23

FIGURE 2.6: MLP SHOWING CONNECTIONS BETWEEN DIFFERENT LAYERS 25

FIGURE 2.7: RNN SHOWING FEEDBACK FROM FIRST OUTPUT OF FIRST HIDDEN LAYER 25

FIGURE 4.1: HEAT TRANSFER FROM THE BODY TO SURROUNDINGS 30

FIGURE 4.2: VELOCITY PROFILE IN A ROOM FITTED WITH A CEILING FAN 32

FIGURE 4.3: MEAN RADIANT TEMPERATURE (MRT) 33

FIGURE 4.4: VICTOR OLGAYAY BIOCLIMATIC CHART IN METRIC UNITS 34

FIGURE 4.5: RELATION BETWEEN PMV AND PPD 35

FIGURE 4.6: ASHRAE STANDARD FOR ADAPTIVE THERMAL COMFORT 36

FIGURE 4.7: CALCULATION OF ANGLE FACTORS 38

FIGURE 5.1: SOLAR RADIATION AND SURFACE TILT ANGLES 43

FIGURE 5.2: RADIATION FLUX ON WALLS AND ROOF FOR A DAY IN THE MONTH OF APRIL. 44

FIGURE 5.3: ANNUAL SURFACE RADIATION FOR DIFFERENT L/W RATIOS 45

FIGURE 5.4: ANNUAL SOLAR RADIATION FOR DIFFERENT BUILDING ORIENTATIONS. 46

FIGURE 5.5: BASIC PRINCIPLE OF EAHE 49

FIGURE 5.6: ANNUAL VARIATION OF GROUND TEMPERATURE AT COIMBATORE 50

FIGURE 5.7: TRIAL PIT WITH THREE TEMPERATURE SENSORS 52

FIGURE 5.8: DIURNAL TEMPERATURE VARIATIONS (FEBRUARY 2016) AT DIFFERENT DEPTHS 52

FIGURE 5.9: MEAN TEMPERATURE VARIATION AT 4FT DEPTH FROM JANUARY TO APRIL 53

FIGURE 5.10: PERFORMANCE OF EAHE DURING SUMMER (MAY 2016) 54

FIGURE 6.1: AIR FLOW RATES FOR WIND ONLY AND BUOYANCY DRIVEN NATURAL VENTILATION 55

Figure 6.2: Airflow patterns around a building without windows 56

Figure 6.3: Ventilation through windward and leeward openings 56

Figure 6.4: Ventilation through windward and side openings 57

Figure 6.5: Variation of wind tunnel data recorded at different wind tunnels 58

Figure 6.6: CFD model used for simulation (all dimensions are in SI units)... 59

Figure 6.7: Effect of Wing walls on natural ventilation for different configurations................................. 60

Figure 6.8: Flow visualization for different Wing wall positions. 62

Figure 6.9: Effect of wing wall length (L) on flow pattern. 63

Figure 6.10: Effect of wing wall length on inlet velocity and construction cost. 63

Figure 6.11: Effect of wing wall height on inlet velocity and construction cost.. 64

Figure 6.12: Front view of Inverted 'L' type of wing wall 64

Figure 7.1: Effect of diurnal variation and thermal mass of building on indoor temperature..................... 67

Figure 7.2: Monthly variation of maximum outdoor temperature, wind velocity and direction.................. 69

Figure 7.3: Diurnal variation of average outdoor temperature and wind velocity in April. 70

Figure 8.1: Effect of controlled nocturnal cooling on indoor temperature 73

Figure 9.1: Simulink model for solving equation 9.1. 78

Figure 9.2: Simulated Indoor temperature for the month of April 79

Figure 9.3: Measured Indoor and outdoor temperatures for the month of April 79

Figure 9.4: Simulink model for training the house model 80

Figure 9.5: Comparison between measured and ANN predicted indoor temperatures 81

Figure 10.1: Indoor temperature without ventilation... 83

Figure 10.2: Indoor temperature with nocturnal cooling.. 84

Figure 10.3: Indoor temperature with both nocturnal cooling and EAHE... 84

Figure 10.4: ANN based predictive Control Law 85

Figure 10.5: Indoor temperature with MLP based ANN (threshold at T_CH) 86

Figure 10.6: Indoor temperature with RNN based ANN (threshold at T_CH) 86

Figure 10.7: Indoor temperature with MLP based ANN (threshold at T_C).. 87

Figure 10.8: Indoor temperature with RNN based ANN (threshold at T_C).. 88

Figure 10.9: Indoor temperature with MLP based ANN (no load from 9:00 to 18:00 with threshold at T_C) 88

Figure 10.10: Indoor temperature with RNN based ANN (no load from 9:00 to 18:00 with threshold at T_C) 89

Figure 10.11: MLP with forced switching off EAHE from 11:00 to 15:00h.. 90

Figure 10.12: RNN with forced switching off EAHE from 11:00 to 15:00h 90

Figure 10.13: MLP with external disturbance (threshold at T-CH) 92

Figure 10.14: RNN with external disturbance (threshold at T-CH) 92

List of Tables

TABLE 2.1: LIST OF ADAPTIVE THERMAL COMFORT MODELS 7

TABLE 2.2: COMFORT TEMPERATURE RECOMMENDED BY CEPT 8

TABLE 2.3: EMPIRICAL MODELS TO PREDICT NATURAL VENTILATION 11

TABLE 2.4: PERFORMANCE OF EAHE IN INDIA 14

TABLE 2.5: PREDICTION OF MICROCLIMATIC CONDITIONS USING ANN 26

TABLE 4.1: HEAT GENERATED BY DIFFERENT ACTIVITIES 30

TABLE 4.2: SCALE FOR RESPONSES TO DIFFERENT THERMAL CONDITIONS 33

TABLE 4.3: WEIGHTAGE FOR CALCULATION OF OPERATIVE TEMPERATURE FOR DIFFERENT VELOCITIES 39

TABLE 4.4: DIFFERENCE BETWEEN OPERATIVE AND INDOOR TEMPERATURES 40

TABLE 5.1: THERMAL PROPERTIES OF BUILDING MATERIALS 46

TABLE 5.2: ENERGY REQUIREMENTS FOR DIFFERENT BUILDING MATERIAL COMBINATIONS 47

TABLE 5.3: ENERGY REQUIREMENT FOR A HYPOTHETICAL ROOF 48

TABLE 6.1: COMPARISON OF VELOCITIES 60

TABLE 6.2: AVERAGE INLET VELOCITY FOR DIFFERENT WING WALL CONFIGURATIONS 61

TABLE 6.3: COMPARISON BETWEEN MARK DEKAY RATING AND CFD RESULTS 61

TABLE 8.1: COMPARISON BETWEEN ESTIMATED AND MEASURED INDOOR MAXIMUM TEMPERATURES 72

TABLE 9.1: AVERAGE DIURNAL DIFFERENCE BETWEEN SOL-AIR AND OUTDOOR TEMPERATURES 76

TABLE 9.2: COMPARISON BETWEEN MEASURED AND SIMULATED TIME LAG AND DECREMENT FACTOR 80

TABLE 10.1: THERMAL PROPERTIES OF BUILDING ELEMENTS USED FOR SIMULATIONS 83

TABLE 10.2: COMPARISON OF VARIOUS PASSIVE COOLING SCHEMES 91

TABLE 10.3: IMPORTANT CLIMATE ZONES OF INDIA 93

TABLE 10.4: COMPARISON OF ENERGY CONSUMPTION AND OPERATIONAL COST 93

TABLE 10.5: ITEM WISE COST FOR THE PROPOSED SYSTEM 94

Acronyms

ACH	–	Air Change per Hour
ANN	–	Artificial Neural Networks
ASHRAE	–	American Society of Heating, Refrigerating and Air-Conditioning Engineers
CEPT	–	Centre for Environmental Planning and Technology
CFD	–	Computational Fluid Dynamics
CHTC	-	Convective Heat Transfer Coefficient
DBT	–	Dry Bulb Temperature
EAHE	–	Earth Air Heat exchanger
EPW	–	Energy Plus Weather data
ET*	–	Equivalent Temperature*
MLP	–	Multi-Layer Perceptron
MPC	–	Model Predictive Control
MRT	–	Mean Radiant Temperature
NC	–	Nocturnal Cooling
NTU	–	Net Transfer Units
OT	–	Operative Temperature
PMV	–	Predicted Mean Value
RH	–	Relative Humidity
RNN	–	Recurrent Neural Network
TERI	–	The Energy and Resources Institute

List of Symbols

A	–	Amplitude of ambient temperature variation [°C]
A_{outlet}	–	Outlet area [m²]
$A_{surface}$	–	External surface area of the building envelope [m²]
C	–	Specific heat of the building material [J/kg K]
C_{air}	–	Specific heat of air [J/kg K]
d	–	damping factor
F_{p-N}	–	Angle factor between the person and the surface
I_{DN}	–	Direct normal solar flux [W/m²]
I_{dH}	–	Clear day diffused solar flux [W/m²]
I_{LW}	–	Intensity of long wave radiation [W/m²]
I_T	–	Total solar radiation consisting of both direct and diffuse solar radiation [W/m²]
k	–	Thermal conductivity [W/m °C]
$\dot{m}_{air}$	–	mass flow rate of air [kg/sec]
$m_{air-room}$	–	Mass of the air inside the room [kg]
Q_{in}	–	Heat generated by occupant and the equipment used inside the building [W]
Q_m	–	Mechanical cooling of the cooling equipment [W]
Q_{solar}	–	Heat transfer across the building envelope [W]
$\left(\dfrac{dQ}{dt}\right)_{gain}$	–	Rate of heat transfer in to the room due to conduction through the walls and roof [J/sec]
$\left(\dfrac{dQ}{dt}\right)_{loss}$	–	Rate of heat transfer out of the room with the help of natural heat sinks (EAHE or nocturnal cooling) [J/s]
R	–	thermal resistance between the external surface and air [m² °C/W]
T_a	–	Ambient temperature [°C]
T_{cool}	–	Outlet temperature of the cooling equipment [°C]
T_CH	–	Upper limit of comfort temperature [°C]

T_CL	–	Lower limit of comfort temperature [°C]
T_C, T_n	–	Comfort temperature [°C]
$T_{exhaust}$	–	Temperature at the outlet of the exhaust fan [°C]
T_{in}	–	Indoor temperature [°C]
T_{in+1}	–	Next hour indoor temperature [°C]
T_M	–	Monthly mean of the outdoor temperature [°C]
T_N	–	Temperature of Nth surface [°C]
T_o	–	Outdoor temperature [°C]
T_{solar}	–	Sol-air temperature on the external surface [°C]
U	–	Transmittance of the building material [W/m² °C]
v_{air}	–	Air velocity at the outlet [m/sec]
w	–	Angular frequency variation [rad/sec]
z	–	Depth of soil [m]
Σ	–	surface tilt angle [degree]
$\forall_{air}$	–	Ventilation rate [m³/sec]
$\forall_{room}$	–	Volume of the room [m³]
α	–	Absorptance
ε	–	Emittance
θ	–	Angle of incidence [degree]
ρ	–	Density of the building material [kg/m³]

Abstract

Energy consumption towards household air conditioning has increased due to global warming and increased income levels. Large scale use of air conditioners may not be sustainable in developing tropical countries and lower strata of the society, which cannot afford air conditioning, will be affected more.

Passive cooling methods as an alternative to conventional air conditioners have been studied in the past. But they have not found wide acceptance in households, mainly due to non-availability of easily measurable thermal comfort standards and lack of integration of different passive cooling methods.

Hence there was a strong motivation to develop a low cost, energy efficient system as an alternative to conventional air conditioners, for use in tropical countries.

To reduce energy consumption for cooling in households, an easily implementable thermal comfort standard was identified. An integrated system, for reducing the heat transfer in to the building, improvement in natural ventilation using external projections called wing-walls, nocturnal (night time) cooling and an innovative low cost Earth Air Heat Exchanger (EAHE) was developed.

Further to improve the performance of the system, a control law with the help of Artificial Neural Network (ANN) was developed. Simulations were initially carried out for a location in Southern India (Coimbatore) and then for different climatic zones in the country.

The control law was found to be effective in reducing the energy consumption towards cooling for all tropical climatic zones and hence was found to be applicable for all tropical countries.

The proposed system resulted in a low-cost cooling system compared to conventional air-conditioners. The installation and operating cost of the proposed system were 60% and 70% lower compared to that of a conventional air-conditioner.

It is felt that wide spread use of the proposed system will not only reduce energy consumption towards cooling in households, but also make it affordable to all sections of society.

Chapter 1

Introduction

Motivation: In tropical countries, where the ambient temperatures are higher compared to temperate regions, thermal comfort is achieved with the help of air conditioners. Thermal comfort is defined as **"that state of mind which expresses satisfaction with the thermal environment"**[1] i.e. the occupants neither feel hot or cold. Air conditioners consumed large amount of electrical energy which was mostly produced by burning fossil fuels. Generation of electrical energy has contributed towards global warming. This has resulted in a higher energy consumption towards cooling, thereby leading to a vicious cycle.

It is felt that this practice of using conventional air conditioners to meet our cooling requirements may not be sustainable in the near future and lower strata of the society, which cannot afford air conditioning, will be affected more. Hence there was a strong motivation to develop a low cost, energy efficient system as an alternative to conventional air conditioners, for use in tropical countries.

Different models have been proposed to quantify thermal comfort. Fanger[2] thermal comfort model has been widely used for air conditioned buildings. It was stated that his comfort model was applicable to all human beings, irrespective of their geographical location and has recommended an indoor temperature of 23 to 27°C.

Subsequently many researchers have proved with help of field experiments, that thermal comfort was a strong function of geographical location. These experiments led to an **Adaptive thermal**

comfort model[3], based on the fact that "if a change occurs such as to produce thermal discomfort, people react in ways which tend to restore their comfort".

The recommended adaptive thermal comfort models were difficult to implement and have not found wide acceptance for use in households.

During the energy crisis in mid-seventies, passive cooling techniques were proposed as an alternative to conventional air conditioners. Passive cooling was classified in to two methods: heat avoidance and heat removal.

Heat avoidance methods included thick and low-density walls (material thermal property), shading, and external surface colour of the building and even physical shifting of occupants. Heat removal, on the other hand, depended on natural heat sinks like sky (radiation), air (convection) and earth (conduction). Due to various reasons, these passive cooling methods have not become popular. Further there has been no attempt to integrate different passive cooling methods.

Passive cooling methods also required manual intervention by the occupants. Given present-day busy lifestyle and high expectations, there is few evidence of such initiatives having been taken by the occupants. Hence there has been a need to automate the whole process.

1.1 Outline of the Thesis

Chapter 2 is an extensive review of the literature on passive cooling methods like building material thermal properties, natural ventilation and Earth Air Heat Exchangers (EAHE). Existing thermal comfort standards, building heat transfer models and utilisation of Artificial Neural Networks (ANN) in controlling indoor temperature have been studied. Gaps identified in 'state of art' have been listed as Objectives in Chapter 3.

Chapter 4 deals with the physiological basis for thermal comfort, limitation of standards and identification of a suitable standard for passively cooled buildings. Simulations and experiments carried out on passive cooling methods formed Chapter 5.

Wind driven natural ventilation was improved using wing walls, which were projections on the building external surfaces, near windows. Results on simulations on effect of wing walls are placed at Chapter 6.

Charter 7: With the help of physical measurements, it was observed that natural ventilation was effective only during nights (nocturnal cooling) and had many barriers.

Chapter 8 discusses ways to mitigate the effect of such barriers. Reasons for lack of popularity of a controlled nocturnal cooling method are identified.

To overcome limitations of existing building heat transfer models, new analytical and ANN models are proposed in Chapter 9.

The proposed model is used for developing a 'control law' in Chapter 10. It is observed that the control law was effective in achieving thermally comfortable indoor environments in tropical countries, without manual intervention, at a much lesser cost compared to conventional air conditioners. The flow chart indicating various aspects dealt in the study along with their inter-dependence is as shown in Figure 1.1 below

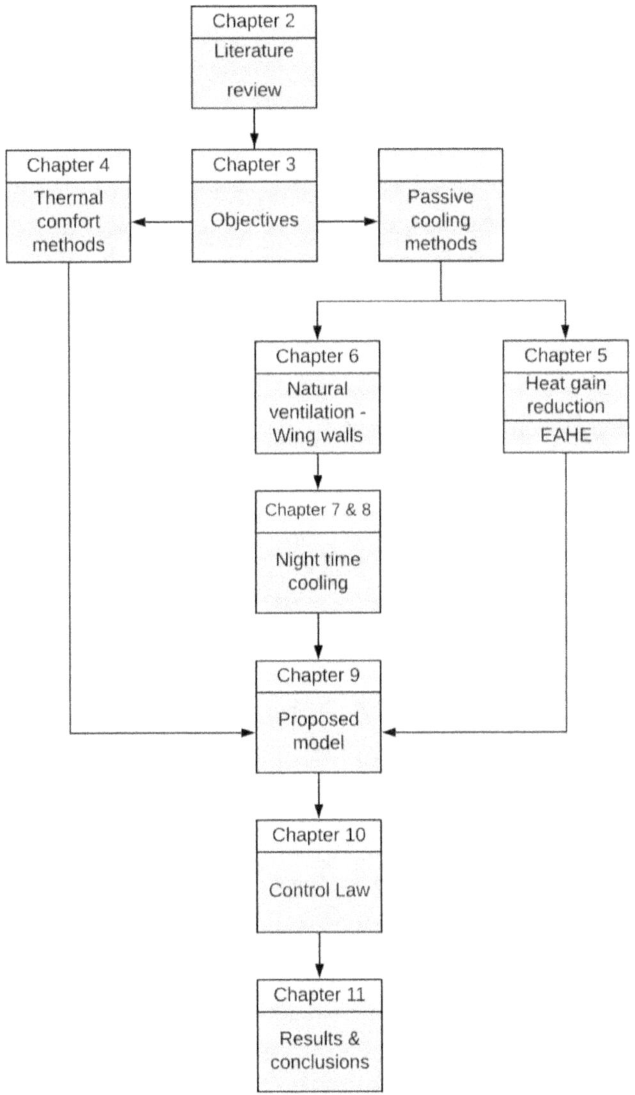

Figure 1.1 Flow chart of thesis

Chapter 2

Literature review

Thermal comfort, for occupants in buildings, is considered more important compared to visual and acoustic comforts. Thermal comfort is **"that state of mind which expresses satisfaction with the thermal environment"**[1].

2.1 Thermal comfort Models

Two types of thermal comfort models, the heat balance model and the adaptive model, are used for air-conditioned and naturally ventilated buildings respectively. The heat balance model was based on the fact that, to experience a thermally comfortable environment, the metabolic heat production in the human body should be equal to the heat lost to the environment. Fanger[2], formulated an empirical formula to determine the thermal comfort levels, based on experiments carried out on persons (young Americans and Europeans) confined in climate chambers. It helped the air conditioning industry to ascertain whether people will feel comfortable in a given thermal environment. Based on this model various standards like ASHRAE Standard-55 and ISO-7730[4] were created. These standards recommended that the summer indoor temperature (T_{in}) should be maintained within a narrow range of 23°C to 27°C (50% Relative Humidity), irrespective of outdoor conditions.

The indoor temperatures of houses in tropical countries usually exceeded the recommended range due to high outdoor temperatures. To maintain the recommended comfortable indoor temperatures, conventional air conditioners have been used for cooling. Further it has also become a practice to set household air conditioner thermostats for an indoor temperature (T_{in}) of $22.5 \pm$ 1°C, irrespective of outdoor temperature (T_o) [5]. For occupants who are used to such comfort levels, the air conditioner is ON throughout the day (and year). Such continuous use of air conditioners has resulted in high energy consumption. This continuous use has been showing an increasing trend in the last few decades[6][7][8].

In order to reduce energy consumption for cooling buildings, designers have reduced infiltration (unintentional air flow) by sealing the building. Such measures have only increased the dependency on air conditioners to achieve thermal comfort.

Many researchers have reported that the Fanger[2] model recommended a lower value for T_{in}, in hot conditions, than what is required. Their efforts, based on field surveys has resulted in an Adaptive thermal comfort model[3] which is based on the fact that **"if a change occurs so as to produce thermal discomfort, people react in ways that tend to restore their comfort."**

It was observed that these adaptive reactions included seasonal clothing, opening appropriate windows, use of external shading structures and even shifting to a comfortable room in the same house depending on the time of the day and the season[9]. Further it was also observed that these reactions were learnt over a period of time (sometimes generations) at any given geographical location.

Details of different adaptive thermal comfort models developed are summarised in Table 2.1 below. Unlike a constant temperature recommended by Fanger[2], the Adaptive model recommended a comfort temperature (T_n) which depended on the mean outdoor temperature of the geographical location. Of the models listed below, the one formulated by Nichol[14] was found to be relevant, as it was based on field experiments carried out in the Indian subcontinent.

Researcher	Period	Equation for neutral temperature	Parameters to be measured
Humpherys[10]	1975	$T_n = 02.56 + 0.830T_i$	Monthly mean of indoor air temperature, T_i
Humpherys[11]	1976	$T_n = 11.90 + 0.534T_m$	Outdoor mean temperature, T_m
Auliciems[12]	1981	$T_n = 17.60 + 0.310T_m$	Outdoor mean temperature, T_m
Griffith[13]	1990	$T_n = 12.10 + 0.530T_m$	Outdoor mean temperature, T_m
Nichol[14]	**1996**	$\boldsymbol{T_n = 17.00 + 0.380T_m}$	**Outdoor mean temperature, $\boldsymbol{T_m}$**
De Dear[15]	1997	$T_n = 17.80 + 0.310T_m$	Outdoor mean temperature, T_m
ASHRAE RP-884[16]	1997	$T_n = 18.9 + 0.225ET^*$	Outdoor mean temperature, ET*
Nichol & Humpherys[3]	2002	$T_n = 13.50 + 0.540T_o$	Outdoor mean temperature, T_o
Nichol & Humpherys[17]	2010	$T_n = 18.80 + 0.330T_{rm}$	Running mean outdoor temperature, T_{rm}
Nguyen[18]	2012	$T_n = 18.83 + 0.34T_m$	Outdoor mean temperature, T_m

Table 2.1: List of Adaptive thermal comfort models

Adaptive thermal comfort models are more realistic because they were based on field surveys conducted in different geographical locations and not on experiments conducted in climate chambers. After extensive validation, the adaptive model, was included in ASHRAE Standard (2013 version), for naturally ventilated buildings.

This adaptive standard recommended measurement of Mean Radiant Temperature (MRT) of the indoor space, in order to determine the thermal comfort temperature. Measurement of MRT required expensive sensors and laboratory equipment, hence this standard has not been used in field conditions.

2.1.1 Indian thermal comfort standards

Bureau of Indian Standards (BIS) 2005[19] recommended an indoor temperature of 23°C-26°C for air conditioned buildings. This standard has been criticised by Centre for Environmental Planning and Technology (CEPT) 2014 report[5]. It was found that thermostats are set at $22.5 \pm 1°C$ throughout the year, in Indian households. Based on field surveys in India, for outdoor temperatures ranging from 16°C-38°C, the report has recommended the following thermal comfort temperatures (Table2.2). It was observed that simple initiatives, like adjustment of the conventional air conditioner thermostat setting to temperatures recommended by CEPT, resulted in some reduction of energy consumption.

Air conditioned rooms	22 °C to 27 °C
Naturally ventilated rooms	20.5 °C to 28.5 °C ± 2.4 °C

Table 2.2: Comfort temperature recommended by CEPT[5]

2.2   Passive Cooling

Passive cooling of buildings by natural ventilation has been an age old strategy. Large floor-to-ceiling heights, orientation of doors and windows with respect to the direction of the mean outside wind etc., have been integral to the design of buildings for a long time. Passive cooling for reducing energy consumption has been a relatively recent phenomenon.

Energy consumption can be reduced by adopting two passive cooling strategies - (a) Reduction of heat transfer in to the building and (b) Removal of indoor heat using natural heat sinks like outdoor air (ventilation) and soil temperature at a depth of more than 2-4m (Earth Air Heat Exchangers).

Extensive research has been done on both the passive cooling strategies, in the last five decades. Important contributions in this area are discussed below

Olgyay[20], used principles from biology, engineering, meteorology, and physics to formulate sustainable building design methodologies. His work has been air flow around buildings. A simple-to-use thermal comfort chart was also introduced by him.

Koenigsberger[21], a German architect, mainly worked in the urban development planning. His works in India are examples of passive cooling strategies. Periodic heat flow equation for a building surface formulated by him was used for developing analytical models.

Givoni[22] worked on passive cooling strategies for more than three decades, and has been acclaimed as a practical resource for studying the impact of basic architectural choices on cooling efficiency. Also his pioneering work on soil conditioning has been further extended.

Mat Santamouris[23] described various state-of-the-art passive cooling techniques, also detailed the limitations of existing passive cooling strategies.

Above mentioned works[20-23] dealt with reduction in heat transferred in to a building. By choosing an appropriate orientation, plan and construction materials, Sushil Kumar et al. [24], studied reduction in energy consumption, by adopting such measures.

2.2.1 Natural ventilation

Natural ventilation reduces the energy consumption towards cooling by replacing the existing hot indoor air by cooler outdoor air. The replacement is carried out by utilising the pressure difference between external and internal surfaces of the building envelope. Such measures were popular in ancient Persian architecture[25].

The pressure difference could be either due to prevailing external winds or accumulation of hot air below the ceiling (buoyancy effect). For present generation buildings, buoyancy effects are negligible and only the wind driven pressure differences are considered[26].

Performance of natural ventilation has been specified either by global mass flow rate or indoor air velocity[26]. Estimation of **mass flow rate** was required for calculating the cooling potential of natural ventilation. On the other hand **indoor air velocity** was used for estimating the increase in comfort temperature levels due to convective heat transfer from the skin surface.

Both these parameters could be calculated by solving the Navier-Stokes[26] equation, under specific boundary and initial conditions. However due to chaotic nature of wind[27], reliable data on initial and boundary conditions, under field conditions is limited. Hence solution of the Navier-Stokes equation has not been a viable option for estimating the performance of natural ventilation.

In the last five decades four different classes of modelling natural ventilation has been developed[28]. They are classified as empirical, network, zonal and CFD models. Empirical and CFD models were found to be useful when dealing with single room houses. Network and zonal models are useful for houses with multiple rooms.

Empirical Models: These models were simple and were used as a first estimate for designing natural ventilation inside a single room. Models which were found to be useful for this study are summarised in Table 2.3 below. Out of the models listed above, only the British Standard BS5925[30] required a few easily measurable input parameters.

Model	Year	Parameter	Remarks
Aynsley[29]	1977	Mass flow rate	Order of Magnitude accuracy for cross ventilation. Good reference for flow visualisation inside a building
British standard BS 5925[30]	**1980**	**Mass flow rate**	**Both cross and single sided ventilation. Also includes buoyancy effects**
De Gidds[31]	1982	Mass flow rate	Limited to single sided ventilation
Givoni B[32]	1965	Indoor air velocity	Restricted to square floor plan
Melaragno[33]	1982	Indoor air velocity	Very useful. No restriction on position of upwind and downwind openings. Wind direction both perpendicular and oblique to the upwind opening
CSTB[34]	1992	Indoor air velocity	Includes effect of site, building orientation, external building characteristics like roof shape and interior partitions. Based on wind tunnel experiments
Ernest[35]	1991	Indoor air velocity	Includes models for different geometries like rectangular, L, U and Z.

Table 2.3: Empirical models to predict natural ventilation.

The mass flow rate models required estimation of pressure coefficient (C_P) on the upwind and downwind surfaces. An empirical formula was developed by Swami[36] in late eighties. This formula was further refined by Grosso[37] in the early nineties. While the Swami's formula predicted surface averaged values of C_P, Grosso's formula yielded an estimate of C_P at any point on the building external surface.

In Coimbatore the monthly mean wind velocities varies from 1m/s to 5m/s[38]. The wind direction is from the South-west direction during the period April to September and reverses to South-east during other months. On any given day wind velocity is high during afternoon and is least during early morning. During summer months the wind velocity reaches very low values resulting in less pressure difference between the windward and leeward sides of the building. Under such conditions natural ventilation will not be effective.

Wing-walls are external projections, (near windows on building walls), which improve the natural ventilation potential of the building under low outdoor wind velocity. Research[39] carried out on wing walls was restricted to study the effect of wing walls on air flow patterns inside the room.

Sushil Kumar et al.[40] carried out a parametric study on wing walls, to quantify their effect on improvement in natural ventilation.

Natural ventilation could be used for cooling, **only if the outdoor temperature (T_o), is less than that of the indoor temperature (T_{in}).** A survey paper[41], concluded that natural ventilation was unpredictable and depended heavily on outdoor weather conditions and that natural ventilation alone was not sufficient to achieve indoor thermal comfort, at all geographical locations.

Natural ventilation is effective for night time cooling[42] in tropical countries. It is a method of passive cooling by which the heat gained inside the building during mornings is released during nights with the help of cooler outdoor air. It helps in reducing the average daily indoor temperature and hence reduces energy consumption. In spite of the cooling potential of night ventilation, there exist[43] several barriers, like reluctance of occupants to open windows during nights, fear of intrusion of insects etc.

2.2.2 Controlled night time cooling

In order to have a better control, Pablo La Roche[44] studied the effect of switching ON, a whole house exhaust fan, whenever the outside temperature (T_o), was lower than the indoor temperature (T_{in}), to reduce hot discomfort hours. Instead of a constant T_{in} of 23°C, a temperature range specified between 'comfort high' (T_{CH}) and 'comfort low' (T_{CL}) was used. Two identical buildings, were studied. Only one had a whole house exhaust fan for nocturnal ventilation. The exhaust fan was switched ON only if $T_{in} > T_o$ and $T_{CL} < T_o < T_{CH}$. A 17% reduction in hot discomfort hours was reported. Simulations also revealed a reduction in energy consumption (average 24%) in only 11 out of 16 different geographical locations.

2.2.3 Earth Air Heat exchanger (EAHE)

EAHE system consists of a pipe buried at a certain depth below the ground surface. At a depth of about 4m, the soil temperature is found to be equal to annual mean outdoor temperature of that location[45]. Outdoor air is circulated through the pipe. A reduction in pipe outlet temperature is observed due to convective heat transfer between the air and inner surface of the pipe. Such a system has been used for indoor cooling.

It has been reported[46] that before 3000 BCE, in Iran, EAHE was used along with tall towers called wind catchers. Heights of these wind catchers ranged between 10m to 20m and were capable of diverting prevailing wind in to underground tunnels for cooling. The cooled air from the tunnels was then used for reducing the indoor temperature of the building. Since construction of tall towers is costly, they have been replaced by centrifugal blowers in present generation EAHEs.

Research carried out on EAHE, in last fifty years, was summarised in a survey paper by Hong Liu[47]. It was observed that most of the research in this area has been carried out in India and European countries. EAHE systems were used only for cooling in India, though in European countries they have been used both for heating and cooling. Use of such systems has shown an increasing trend in European countries.

Important EAHE research carried out in India along with their salient features are listed in Table 2.4.

Researcher	Year	Location	Depth (m)	Pipe length (m)	Temperature reduction (°C) $T_{out} - T_{outlet}$
Bansal NK[48]	1986	Mathura	1.5	1000	19
Girija S[49]	1999	Gujarat	3	50	14
Thanu NM[50]	2001	Delhi	4	76.5	12
Ashih S[51]	2005	Delhi	1.5	78	5.9
Ashih S[52]	2008	Delhi	1.5	78	3°C reduction indoor temperature
Rohit M[53]	2013	Rajasthan	3.7	60	Hybrid system to cool air conditioner inlet

Table 2.4: Performance of EAHE in India

2.2.3.1 Summary

- Most of the work carried out on EAHE was restricted to North India, which has a large variation in diurnal temperature.

- Pipes were laid at a depth of 1.5 to 4m.

- Reduction of pipe outlet temperature when compared to ambient temperature (T_{out}), has been reported in all projects.

- Only one researcher has reported reduction in indoor temperature with the help of EAHE.

- Transient performance of EAHE has not been studied.

- Installation was carried out for large buildings and not small households.

Ajmi[54] carried out simulations for hot-arid desert climate of Kuwait. He used a 60m long pipe buried at a depth of 2m. His simulations indicated a reduction in indoor temperature by 2.8°C during peak summer.

Vikas Bansal[55] recently studied transient behaviour of EAHE during continuous operation of up to 36 hours. The work was based on CFD analysis of a 100m long pipe. Deterioration of EAHE operation during continuous operation was reported.

In spite of persistent efforts by researchers in the last five decades, EAHE has not been a popular choice for households, primarily because of cost of a 1.5-4m deep trench outside the house. Out of the 41 projects listed as most energy efficient buildings in India, by The Energy and Resources Institute (TERI)[56], only 5 projects used EAHE for indoor cooling.

2.3   Indoor temperature model

Passive cooling (mentioned in previous paragraphs), was based on a common assumption that an equivalent temperature (Sol-air) on the building envelope's outer surface can be measured. Measuring Sol-air temperature had its own limitations[57] and was difficult for estimating thermal performance of buildings.

The indoor temperature of a room is a complex function of several parameters such as weather, building material, house plan and its orientation, etc. Givoni[58] proposed an empirical formula for predicting the indoor maximum temperature with the help of known outdoor mean temperature. The formula was useful only for predicting peak cooling load of the building and did not provide information on hourly variation of indoor temperature.

2.3.1 Analytical building heat transfer models

Following building heat transfer models[21] were widely used in the industry, for estimating the cooling capacity of air conditioning systems.

- Steady State

- Periodic heat flow

- One dimensional transient heat transfer

2.3.1.1 Steady state model

It assumed that steady state conditions prevail in both outdoor and indoor spaces. This assumption was valid for hot-humid climates, where the diurnal change in outdoor temperature was less and the indoor temperature was maintained at constant temperature ($23\,^\circ$C) using conventional air conditioners. The model was based on the following equation.

$$Q_m = Q_{solar} + Q_{in}$$

Eqn. 2.1

Where

Q_m	–	Mechanical cooling of the cooling equipment [W]
Q_{solar}	–	Heat transfer across the building envelope [W]
Q_{in}	–	Heat generated by occupant and the equipment used inside the building [W]

Since Q_{in} was dictated by minimum requirements of the occupants, the aim of the building designer was to reduce Q_{solar} to the extent possible, so that energy required for cooling was reduced.

The cooling potential of the equipment Q_m was given by the following formula.

$$Q_m = 1300 * \forall_{air} * [T_{cool} - T_{indoor}]$$

Eqn. 2.2

Where

$\forall_{air}$	–	Ventilation rate (m^3/s)
T_{cool}	–	Outlet temperature of the cooling equipment ($^\circ$C)
1300	–	Volumetric specific heat of air (J/m$^{3}\,^\circ$C)

The ventilation rate was expressed in terms of Air Changes per Hour (ACH) using the following equation.

$$\forall_{air} = \frac{ACH * Volume\ of\ room}{3600}$$

Eqn. 2.3

Q_{solar} for a surface with transmittance 'U' (walls and roof) was calculated using the following equation.

$$Q_{solar} = A_{surface} U [T_{solar} - T_{indoor}]$$

Eqn. 2.4

where

$A_{surface}$ –   External surface area of the building envelope

U –   Transmittance of the building material

T_{solar} –   Sol-air temperature on the external surface

2.3.1.2 Periodic flow model

The steady state model discussed above was valid only if the building envelope materials did not have any heat storage capacity and diurnal temperature variations were less. However in actual conditions it was not valid, except for warm-humid tropical conditions, where diurnal variation was very low and building materials used are thin and of low densities.

Since the outdoor temperature shows a diurnal variation, part of the heat is stored in the morning hours and released into the indoors during nights. Since this phenomenon is cyclic, a periodic heat flow model was used to describe it. A typical variation of diurnal indoor and outdoor temperature is as shown below in Figure 2.1.

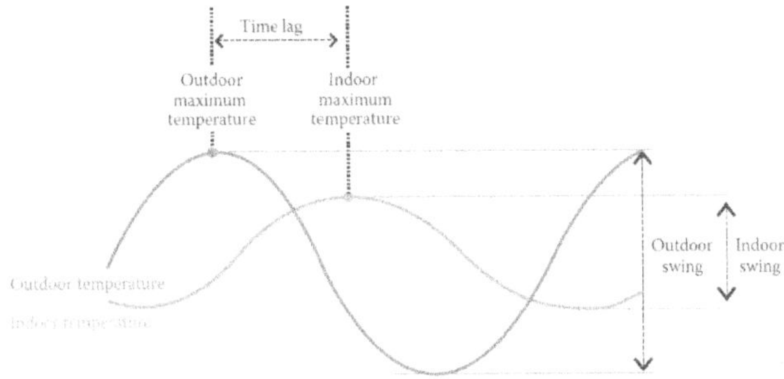

Figure 2.1: Diurnal variation of indoor and outdoor temperatures

Two important parameters were used to define this model.

- Decrement factor $f = \dfrac{T_{indoor\,max}}{T_{outdoor\,max}}$

- Time lag $\varphi(h)$: the time difference between occurrence of $T_{o\,max}$ and $T_{i\,max}$

Both these parameters depend on thermal diffusivity (α) and transmittance (U) where

$$\alpha = \frac{k}{\rho C} = \frac{Conductivity}{Volumetric\ specific\ heat}$$ Eqn. 2.5

Where

k —   Thermal conductivity

ρ —   Density of the building material

C —   Specific heat of the building material

Decrement factor and time lag as a function of thickness and density were determined experimentally[59] and is reproduced below.

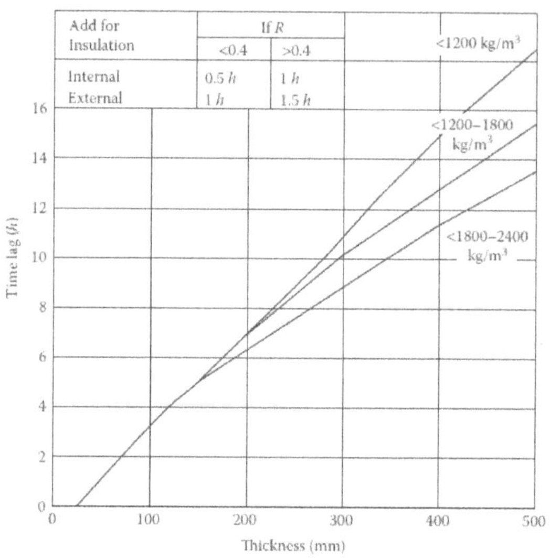

Figure 2.2: Variation of time lag as a function of thickness and density[59]

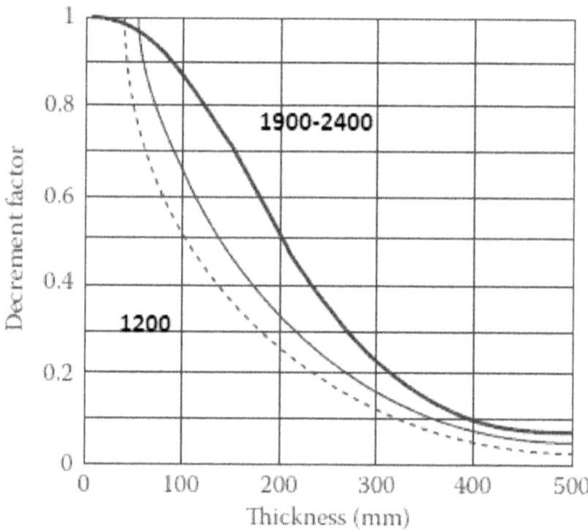

Figure 2.3: Variation of decrement factor as a function of thickness (mm) and density kg/m³ [59]

For a constant indoor temperature, the periodic heat flow was sum of two parts.

Average heat flow during one day

$$Q_{av} = AU(T_m - T_i)$$

Eqn. 2.6

Deviation from average due to f and φ

$$Q_{dev} = AU(T_\varphi - T_m)f$$

Eqn. 2.7

Where T_φ is $T_{Sol-air}$ at Φ hours earlier

$$Q = AU[(T_m - T_i) + f(T_\varphi - T_m)]$$

Eqn. 2.8

Vast database of f and φ has been reported[60][61]. Equation 2.8 was used for estimating periodic heat flow through a building, with a constant indoor temperature.

2.3.1.3 One dimensional transient heat transfer model[62]

This model assumed that heat transfer in to a wall is one dimensional and it was given by following equation.

$$\rho c \frac{dT}{dt} = k \frac{d^2T}{dx^2}$$

Eqn. 2.9

The boundary conditions, at the exterior and internal surfaces were given by following.

$$\left(-k\frac{dT}{dx}\right)_{x=0} = h_{out}(T_{sa} - T_{out})$$

$$\left(-k\frac{dT}{dx}\right)_{x=L} = h_{in}(T_{s,i} - T_{in})$$

Eqn. 2.10

Where h_{out} and h_{in} are convective heat transfer coefficients at the outer and inner wall surfaces respectively. $T_{s,i}$ and T_{sa} are temperatures at the inner surface and Sol-air temperature at the external surface of the wall.

The temperature at time $t = 0$, was taken as the initial condition. For the given boundary and initial conditions, the solution for Equation 9.9 was of the form given below [63].

$$T(x,t) = \sqrt{X_1^2 + X_2^2}[\sin(wt + \varphi)]$$

Eqn. 2.11

From above equation the time lag (φ) and decrement factor were calculated by following equations.

$$\varphi = \arctan\left(\frac{X_2}{X_1}\right)$$

Eqn. 2.12

$$f = \sqrt{X_1^2 + X_2^2}$$

Eqn. 2.13

2.3.1.4 Limitations of models

All three models discussed above required accurate measurement of Sol-Air temperature (T_{Solair}). Indoor temperature was assumed to be at a constant value given by the Fanger model, instead of comfort range recommended by the "Adaptive thermal comfort" model.

2.3.1.5 Sol-Air temperature[57]

It was defined as the outside air temperature which, in the absence of solar radiation, would give the same temperature distribution and rate of heat transfer through a wall (or roof), due to the combined effects of the actual outdoor temperature and the incident solar radiation. It was given by the following formula[57]:

$$T_{solair} = T_{out} + R(\alpha I_T - \varepsilon I_{LW})$$

Eqn. 2.14

Where

R	–	thermal resistance between the external surface and air [m^2 °C/W]
α	–	Absorptance
ε	–	Emittance
I_T	–	Total solar radiation consisting of both direct and diffuse solar radiation
I_{LW}	–	Intensity of long wave radiation from a black body at the given outdoor temperature [W/m^2]

ASHRAE had recommended following values for above mentioned parameters:

R — 0.05 and 0.04m^2 °C/W for walls and roof respectively

α — 0.9 and 0.5 for dark and light coloured surfaces respectively

ε — 0.9 for light coloured surface

I_{LW} — 0 and 100W/m^2 for walls and roof.

Thermal resistance (R): It is the reciprocal of surface heat transfer coefficient, which in turn depends on the Nusselt number (calculated using empirical formulae). Though ASHRAE recommended 'R' values, for ease of calculation, it showed a large variation with wind speed as shown in Figure 2.4 below. Hence 'R' was difficult to predict accurately, under field conditions.

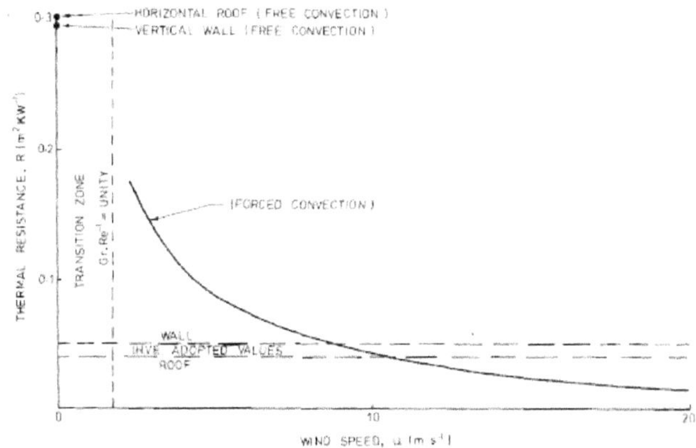

Figure 2.4: Variation of thermal resistance with wind speed [57]

2.4 Artificial Neural Networks (ANN)[63]

In order to overcome the limitations of the analytical models, several researchers have used Artificial Neural Networks (ANN) to model ambient weather conditions around a building and its indoor temperature.

The behaviour of ANNs was similar to the human brain. It followed a two stage process of acquired knowledge (training) and then retained it for reacting to familiar inputs. ANN based model was also referred to as 'Black box' model, because it did not require detailed information like analytical models. Hence development time for ANN models was comparatively less.

Artificial neurons called perceptrons were fundamental building blocks for ANN. A perceptron (Figure 2.5) took several inputs x_1, x_2 ... x_n and gave a single value output. A simple rule to compute the output was by introducing weights, w_1, w_2... which were real numbers, expressing the importance of the respective input to the output. The perceptron's output, zero or one, was determined by whether the weighted sum is greater than some threshold value (Activation function).

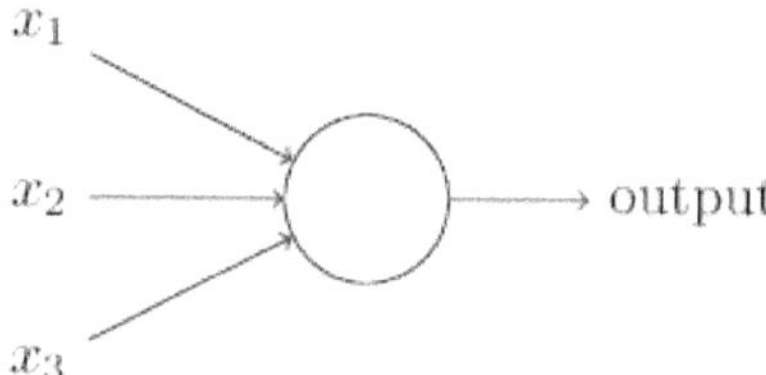

Figure 2.5: Perceptron with three inputs and one output[64]

Mathematically perceptron was expressed as a function $P(x)$.

$$P(x) = \begin{cases} 0, & if \sum_{i=1}^{n} x_i w_i < threshold \\ 1, & if \sum_{i=1}^{n} x_i w_i \geq threshold \end{cases}$$

Eqn. 2.15

Where

x	—	a vector of dimension n
n	—	number of inputs to neuron
x_i	—	i^{th} element of x
w_i	—	weight corresponding to i^{th} element in x (i.e. x_i)

A more common way of representing $P(\mathbf{x})$ was to introduce a bias (b) as shown below.

$$P(\mathbf{x}) = \begin{cases} 0, & if \ \sum_{i=1}^{n} x_i w_i + b < 0 \\ 1, & if \ \sum_{i=1}^{n} x_i w_i + b \geq 0 \end{cases} \qquad \text{Eqn. 2.16}$$

Values of biases and weights were varied to model decision-making. A perceptron alone was not very powerful and could not model even a XOR logic gate. However, a complex network of perceptrons was found to be more powerful in decision-making.

Multi-Layer Perceptron (MLP):

Multilayer Perceptron, shown in Figure 2.6, was built using multiple layers, where each layer consisted of many perceptrons. The left and right most layers are called input and output layers respectively. All other layers are called hidden layers. Output of each perceptron in a given hidden or input layers was connected to all perceptrons in the next layer. Hidden layers are used to learn the mapping of inputs to outputs, for different training data sets.

Recurrent Neural Network (RNN):

The RNN was an ANN with at least two hidden layers, with feedback from the first-layer output as shown in Figure 2.7. RNNs were found to be more effective in responding to temporal changes in the input. This was achieved by storing past outputs in a delay element (**D**) and giving weighted feedback to the first layer activation function. RNNs were found to be relatively difficult to train when compared to MLP.

Though the concept of ANN was developed much earlier, it was used for predicting energy performance of buildings, only in the late nineties by Nannaririello[64]. A. Kalogirou[65] used an ANN for predicting energy consumption (both heating and cooling) in a house situated at latitude of 45°N. Natural ventilation was not used, since the house had no windows. Ruano[66] used an ANN for modelling a constructed house and controlling an air conditioner. ANN model was able to predict indoor temperature thirty minutes in advance. This information was used to switch ON/OFF the air conditioner. For training the ANN, the model relied on a difficult-to-measure parameter like solar radiation. A 27% saving in energy consumed by the air conditioner has been reported.

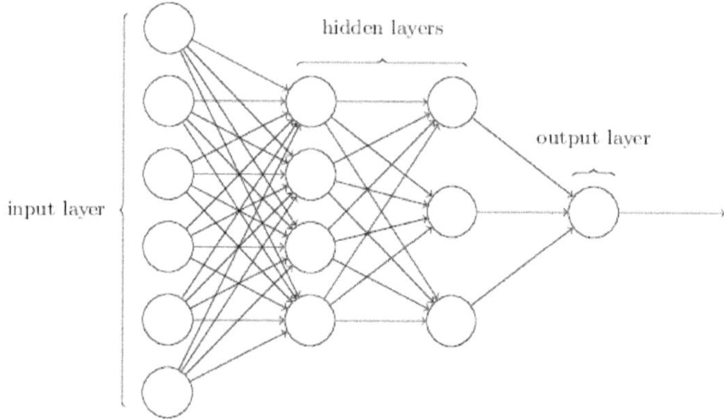

Figure 2.6: MLP showing connections between different layers [63]

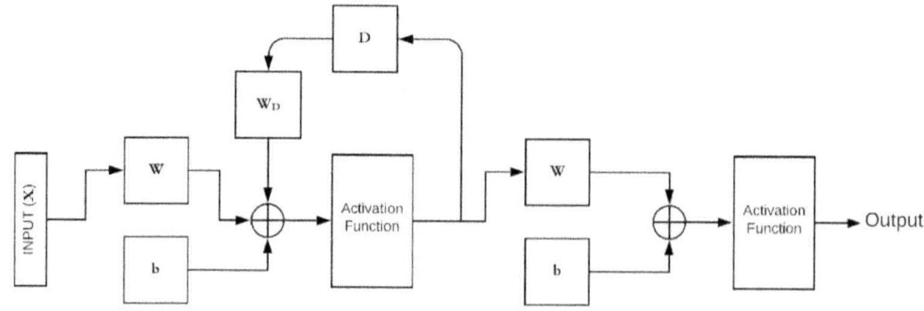

Figure 2.7: RNN showing feedback from first output of first hidden layer [63]

In conventional electronic feedback control systems, the current system output is measured and compared with the desired output, to calculate an error voltage. The error voltage is then used for driving the system in such a way that the error is reduced.

The control scheme used by Ruano[66] is called Model Predictive Control (MPC). It is a control strategy wherein generation of error voltage, is based on both current input and future output of the system. This is achieved by feeding the current input and output of the system to a trained ANN, for generating future output of the plant. MPC became a popular choice in the chemical process industry after Camacho[67] published his work in the year 2003.

Moon[68][69] studied application of ANN based MPC for controlling comfort temperature in conventionally air conditioned buildings. Their simulations indicated that such a scheme prevented undershoots and overshoots but did not reduce energy consumption towards cooling.

Several researchers have used ANN for predicting microclimatic conditions that affect thermal performance of buildings, in locations where climate data was not available. Such ANN models were based on at least one year site specific data. Their efforts are summarised in Table 2.5 below.

Researcher	Year	ANN application
Alawi[70]	1998	Prediction of solar radiation for a given site
Kemmoku[71]	1999	Prediction of solar radiation next day
Kalogirou[72]	1999	Prediction of wind speed data based on 11 years of historical data
Tasadduq[73]	2002	Prediction of ambient temperature 24h in advance for a given location
Reddy[74]	2003	Estimation of hourly solar radiation in India

Table 2.5: Prediction of microclimatic conditions using ANN

Afran[75] in a survey paper summarised research carried on the application of ANN for improving building thermal performance.

2.4.1 Summary

Modelling was based on difficult to measure parameters like solar radiation.

No attempt was made to use an analytical model with building envelope thermal properties, to train the ANN. **Hence the model is site specific and requires at least one year of data.**

No attempt has been made to utilise the potential of ANN for controlling the indoor temperatures of buildings, which used passive cooling techniques like natural ventilation and EAHE. Effect of ANN based control, on achieving adaptive thermal comfort has not been studied.

Chapter 3

Objectives of the Thesis

Based on the research gaps identified in the literature review, this study concentrated on following areas.

- Are existing thermal comfort models relevant for naturally ventilated buildings?

- Can optimization of the building envelope lead to reduction in energy consumption towards cooling?

- Can installation cost of EAHE be reduced?

- Parametric study of wing walls to improve the natural ventilation potential of the building.

- Can Integration of different passive cooling techniques result in achieving comfortable indoor temperatures?

- Is it possible to develop analytical and ANN based models to predict the indoor temperature in passively cooled buildings using easily measurable parameters?

- Is it possible to formulate an ANN aided control law to mimic important aspect of human behaviour i.e. **learning over a period of time** for implementation of adaptive thermal comfort model in its truest sense?

- Demonstrate a low-cost alternative to conventional air conditioners suitable for tropical countries.

Chapter 4

Thermal comfort models

Over a period of time, people living in tropical developing countries have adopted western life styles. This has resulted in artificially maintaining western indoor thermal comfort standards, irrespective of local climatic conditions. Sustaining such artificially maintained conditions, may not be feasible in the long run, due to both global warming and increasing cost of energy. It is in this context, physiological basis of, and different models of, thermal comfort are discussed.

4.1   Physiological basis of thermal comfort

Physiology is a branch of biology, which deals with how organs and systems within body communicate with each other, so that their combined efforts result in favourable conditions of survival. Ecophysiology is a sub branch of physiology that deals with adaptation of human beings to environmental conditions. Thermal comfort is one such environmental condition to which human beings are very sensitive.

Heat generated by Human beings:

Heat is generated in ~~Human~~ the body due to two kinds of biological processes: basal metabolism and muscular metabolism (while carrying out work). Heat generated, while carrying out different activities is given in Table 4.1 below.

Activity	Heat generated (J/s)
Sleeping	72
Reclining	81
Seated	108
Standing	128
Reading	99

Table 4.1: Heat generated by different activities[76]

Human skin temperature is maintained between 31 °C to 34 °C[77], for thermally comfortable conditions. Heat generated by the body during physical activities, is transferred to the environment, through the skin. If external conditions are not suitable for heat transfer, the body temperature increases and results in thermal discomfort.

Heat transferred from human body (due to metabolic rate (M)) to environment is through Conduction (C_d), Convection (C_v), Radiation (R_d) and Evaporation (E), as shown in Figure 4.1 below.

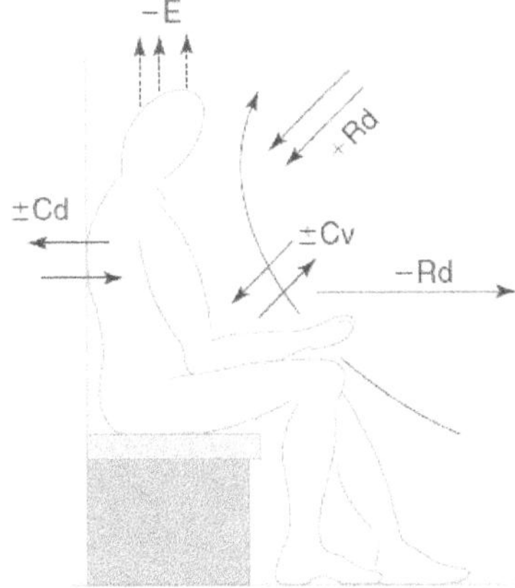

Figure 4.1: Heat transfer from the body to surroundings[77]

The heat balance equation[77] is

$$M \pm Rd \pm Cv \pm Cd - Ev = \Delta S$$

Eqn. 4.1

If change in heat stored (ΔS) within the body is positive, the body temperature increases and causes thermal discomfort. The human body reacts to this condition by sweating, wherein additional heat accumulated is reduced by evaporation. However, prolonged sweating leads to dehydration and in extreme conditions to death. Hence, to avoid such unfavourable conditions, the indoor environment should be such that, required heat transfer takes place from human body to the environment.

4.1.1 Environmental thermal comfort factors

Heat transfer from the human skin to the environment depends on three factors: a) Air temperature, b) Air movement and c) Relative humidity.

Air Temperature:

The temperature of the surrounding air, also called Dry Bulb Temperature (DBT), is measured using a thermometer or temperature sensor. It is the most important factor as it affects the convective heat transfer from the human skin. Further it is the easiest to measure compared to other factors.

Air Movement:

It affects both the coefficient of convective heat transfer, as well as, the evaporation rate. Indoor velocity is usually less than 3 m/s. Due to low velocities encountered inside a room, it is measured by using expensive hot-wire anemometers. Empirical data are used to estimate the indoor air velocity, under field conditions.

It has been recommended that for every 0.275 m/s (restricted to a maximum value of 1.5m/s) increase in indoor air velocity, the upper limit of comfortable temperature increased by 1°C [1]. The upper limit of indoor velocity was subsequently increased to 2 m/s for tropical countries, which corresponded to an increase in comfort temperature limit by 6 °C [22]. Ceiling fans have been used in tropical countries to increase the upper limit of thermal comfort temperature. Velocity profile[78] inside a room fitted with a ceiling fan is as shown in Figure 4.2. It was observed that an air velocity up to 1.5m/s (300ft/min) was achieved easily.

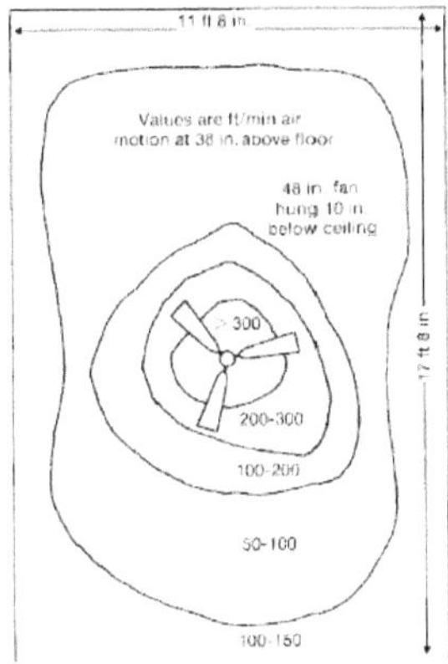

Figure 4.2: Velocity profile in a room fitted with a ceiling fan[78]

Relative Humidity (RH):

It affects evaporation rate from the skin and is measured using relatively cheap humidity sensors, with an accuracy of ± 20%.

Mean Radiation Temperature (MRT):

Heat transfer from human skin to surrounding surfaces (walls, floor and ceilings), through radiation, is measured with MRT. It is defined as that uniform temperature of an imaginary black enclosure (t_r), which would result in the same heat loss by radiation from the person (R) as the actual enclosure (R'). The concept of MRT is as shown in below.

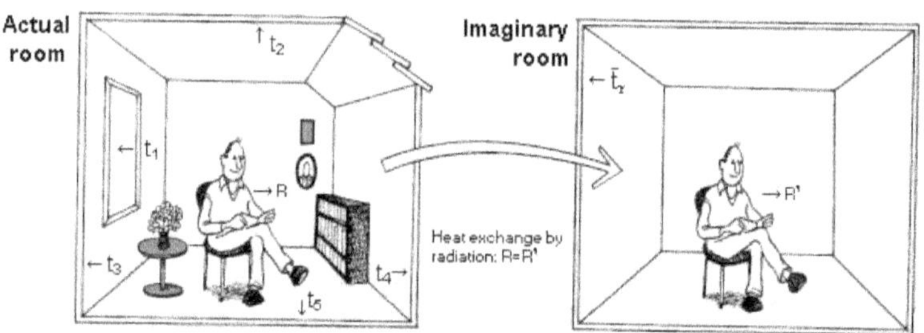

Figure 4.3: Mean Radiant Temperature (MRT)[79]

4.2   Thermal comfort models

ASHRAE defined thermal comfort as "condition of mind that expresses satisfaction with the thermal environment". Obviously it required a subjective evaluation of factors that are not only physiological but also included factors like past experience, culture etc. For achieving thermal comfort, Equation 4.1 alone was not the sufficient condition. Hence questionnaires were used for determining comfort levels. The responses were then classified on a seven point scale as given table (Table 4.2) below.

Scale	Response
3	Hot
2	warm
1	Slightly warm
0	Neutral
-1	Slightly cool
-2	Cool
-3	Cold

Table 4.2: Scale for responses to different thermal conditions[1]

Based on measurements and questionnaires following thermal comfort models were developed.

4.2.1 Victor Olgyay Bioclimatic chart

Victor[20] proposed the bioclimatic chart given below in Figure 4.4, which indicated a combined effect of environmental factors DBT, RH and air velocity, on thermal comfort. The shaded region indicated the thermally comfortable region. His chart was widely used, because of its simplicity. It also recommended a wide range of comfort temperature (20°C to 35°C). It was also evident from the chart that wide range of relative humidity can be tolerated, without causing thermal discomfort. **Hence DBT and air velocity are relatively more important, out of the three factors mentioned above.**

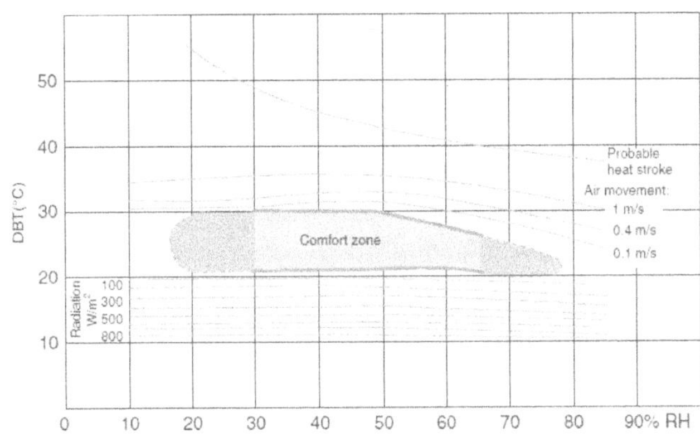

Figure 4.4: Victor Olgayay bioclimatic chart in metric units [20]

4.2.2 Fanger heat balance (comfort) equation[2]

Fanger carried out a series of experiments during mid-seventies on people housed in climate chambers. Subjects were asked to rate the thermal comfort level on a 7 point scale given in Table 4.2 above. Based on heat balance equation, Fanger also coined the term Predicted Mean Value

(PMV), which indicated the scale of response given in Table 4.2 above. PMV was calculated using following inputs.

- Body mass

- Body surface

- Vapour pressure of ambient air

- Air temperature

- Fraction of body clothed

- Surface Temperature of garment

- MRT

- Convection heat transfer coefficient

PMV values were then used to define the Predicted Percentage of Dissatisfied (PPD) people, using Figure 4.5 below. **It was evident from above that PMV can be calculated only under laboratory conditions and was not suitable for field applications.**

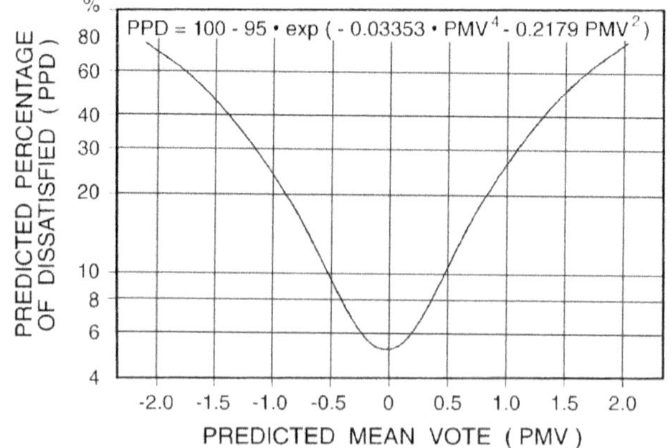

Figure 4.5: Relation between PMV and PPD[2]

4.3 Adaptive thermal comfort model

Fanger comfort model was based on experiments carried out in climate chambers on young Americans and Europeans. He stated that his comfort equation and PMV model was applicable to all human beings, irrespective of their geographical location and climatic conditions. Subsequently many researchers have proved with help of field experiments, that thermal comfort was a strong function of geographical location. These experiments led to the adaptive thermal comfort model. It was based on the fact that "if a change occurs such as to produce thermal discomfort, people react in ways which tend to restore their comfort". Various adaptive thermal comfort models have been developed by researchers [2][10]-[18].

The adaptive thermal comfort model presently recommended by ASHRAE[16] is as given in Figure 4. 6 below. In addition it also indicated 80% and 90% acceptance levels.

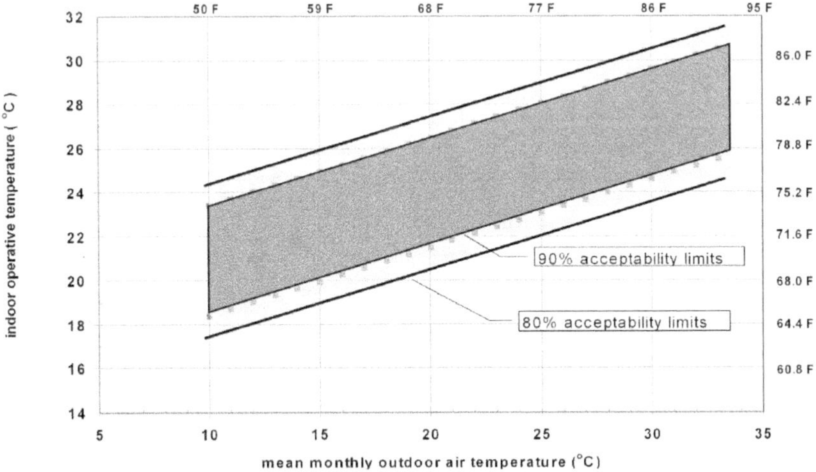

Figure 4.6: ASHRAE standard for adaptive thermal comfort[16]

The ASHRAE adaptive model has been verified by experiments carried out at different geographical locations. It provided a better estimate of comfort temperature, compared to the Fanger model.

The ASHRAE adaptive model specified indoor comfort temperature in terms of an Operative Temperature (OT). It is defined as "the temperature of a uniform isothermal black enclosure in which a person would exchange heat by radiation and convection, at the same rate, as given in the non-uniform environment". It is given by the following weighted average of convection and radiation modes of heat transfer.

$$OT = AT_{in} + (1 - A)MRT$$

Eqn. 4.2

Where "A" takes values depending on the indoor velocity.

Mean Radiant Temperature (MRT): It was calculated using following equation[80].

$$T_R^4 = T_1^4 F_{p-1} + T_2^4 F_{p-2} + \cdots + T_N^4 F_{p-N}$$

Eqn. 4.3

Where

$$T_N \quad - \quad \text{temperature of N}^{th} \text{ surface}$$

$$F_{p-N} \quad - \quad \text{angle factor between the person and the surface N}$$

Calculation of angle factors is carried out using the chart given below in Figure 4.7

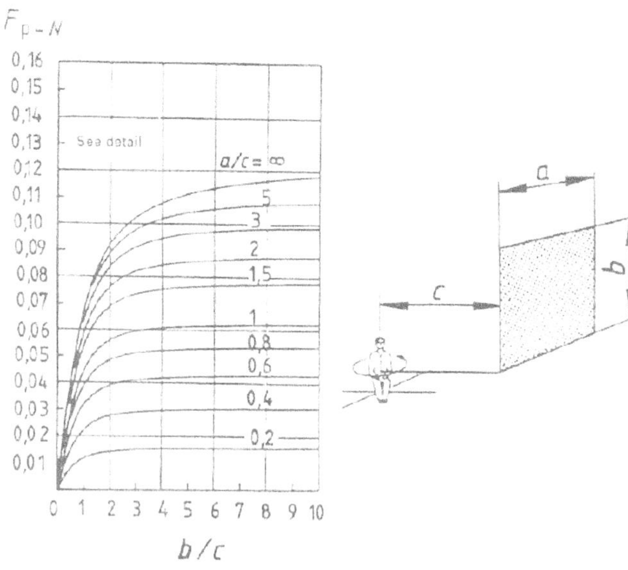

Figure 4.7: Calculation of Angle factors[80]

It was observed From Equation 4.3 above, that calculation of MRT required measurement of all surface temperatures with multiple temperature sensors. Similarly, the angle factors were also difficult to estimate. because it depended on the relative position of the occupant with respect to various surfaces. **Hence, measurement of MRT and in turn OT, was extremely difficult under field conditions.**

Additionally, it has been observed that OT it is not suitable above 27 *°C,* as it neglects evaporation losses[22]. Hence it is not suitable for tropical countries, where the indoor temperature often exceeds 27°C. **Due to above mentioned reasons, the ASHRAE adaptive thermal comfort model has not been used for field conditions.**

4.4   Identification of suitable adaptive thermal comfort model

Due to limitations of both the Fanger and the ASHRAE adaptive models, mentioned above, it was decided to adopt a novel adaptive thermal comfort model, suitable for passively cooled buildings.

ASHRAE had recommended an upper limit (1.5m/s) for indoor velocity for naturally ventilated buildings. Indoor velocities in the order of 2m/s, was observed during physical measurements in Coimbatore. Hence weightage 'A' used in Equation 4.2 was extrapolated for an indoor velocity of 2m/s. The values of "A" chosen for different indoor velocity are given below.

Velocity (m/s)	0.20 to 0.60	0.60 to 1.00	1.50	**2.00**
A	0.60	0.70	0.82	**0.88**

Table 4.3: Weightage for calculation of operative temperature for different velocities

Equation 4.2 was suitably modified for an indoor velocity of 2m/s indoor velocity. The operative temperature is accordingly modified to give:

$$OT = 0.88T_{in} + 0.12MRT$$

Eqn. 4.4

Experiments conducted in a room during summer at Coimbatore indicated that indoor wall temperatures were approximately same as the indoor air temperature. For an indoor temperature of 30°C, only the roof showed a variation of 32°C to 36°C. The room floor area was 3.04m by 3.04m and had a ceiling height of 3.04m.

The operative temperature for the room was calculated, using Equation 4.4 and angle factor chart (Figure 4.7), for a person in sitting posture. The difference between indoor temperature and OT is given in Table 4.4.

T_{roof}	36.00	34.00	33.00	32.00
T_{in}	32.00	31.00	31.00	29.00
MRT	26.85	25.89	25.76	24.25
OT	31.38	30.38	30.37	28.43
OT - T_{in}	**-00.61**	**-00.61**	**-00.62**	**-00.56**

Table 4.4: Difference between Operative and Indoor temperatures

It was observed that for above mentioned building, the difference between OT and indoor temperatures was small. Hence an adaptive model based on indoor temperature is sufficient to determine the comfort temperature. Accordingly the Nicol's model[14], which was based on indoor temperature only, **was identified as most suitable, for naturally ventilated buildings.**

Chapter 5

Passive and Natural Cooling

Passive and Natural Cooling methods are building design techniques, for indoor temperature reduction, using naturally occurring processes.

Passive cooling methods:

- The indoor temperature is reduced by adopting an appropriate building plan (length to width ratio), fixing the building orientation and a choice of construction materials.

- By reducing the amplitude of cyclic indoor temperature variation and by increasing its time delay (phase shift), the heat gain is modulated.

The following natural cooling methods are being used for removal of heat from the building envelope:

- Radiative cooling from roofs by manually closing the shutters over the roof during mornings and removing them during nights (to enhance long wave radiation).

- Evaporative cooling by spreading jute bags over the roof and sprinkling water.

- Using an Earth Air Heat Exchange (EAHE) system to replace hot air accumulated inside the building.

- Ventilation cooling by using a steady flow of cool outdoor air when outdoor temperature is less than indoor temperature.

Radiative and Evaporative cooling have not been considered.

5.1   Heat gain reduction and modulation

Solar radiation on the building envelope leads to transfer of heat into the building, resulting in an increased indoor temperature. In the absence of air conditioning, indoor temperatures in tropical countries exceed acceptable thermal comfort levels.

Efforts have been made using guidelines[81] to minimize the heat transfer by addressing issues like the building footprint, its orientation and thermal properties of building materials.. However, data on percentage reduction of solar radiation on the building envelope and consequent reduction in energy consumed towards air conditioning were not available.

The annual solar radiation on different building surfaces (plinth area of $18.3m^2$ and ceiling height 3.048m) for 365 days was calculated for Coimbatore city ($11.01°N$ latitude and $76.97°$ E longitude) located in South India. Effect of changes in the building footprint on the incident solar radiation on the building was studied and the footprint was optimized. Using this footprint, a study on the building orientation was carried out. Various combinations of building materials for a roof were then studied, to further reduce the effect of incident solar radiation.

5.1.1 Solar Radiation on the Building Envelope

Solar radiation, incident on exterior wall surfaces and roof, consists of 3 components: direct radiation (I_D), diffused radiation (I_d) and reflected radiation (I_r). For estimating the transfer of heat into the building envelope, the reflected radiation component (I_r) is very low and hence was neglected. Diurnal radiation on the wall surfaces and roof was calculated for Coimbatore for all months using following relations[82].

$$I = I_D + I_d$$ Eqn. 5.1

$$I_D = I_{DN}\cos(\theta)$$ Eqn. 5.2

$$I_d = \frac{I_{dH}(1 + \cos(\Sigma))}{2}$$ Eqn. 5.3

Where

I_{DN} –   Direct normal solar flux [W/m²]

I_{dH} –   Average clear day diffused solar flux [W/m²]

θ –   Angle of incidence

Σ –   surface tilt angle (as shown in Figure 5.1)

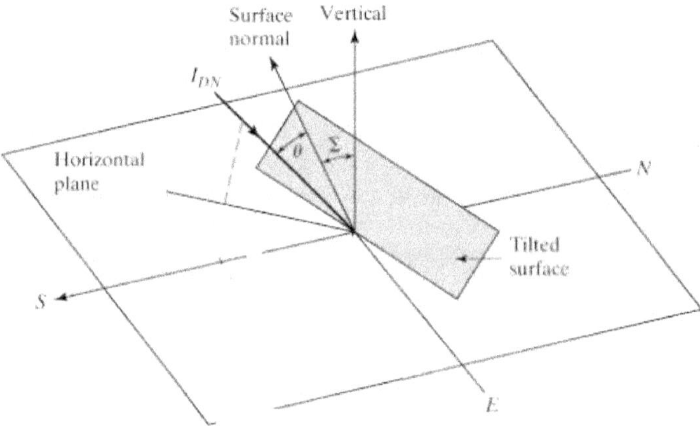

Figure 5.1: Solar radiation and surface tilt angles[82]

Using above equations, it was found that the radiation was maximum during the summer month of April. The calculated diurnal variation of total solar radiation flux on all external surfaces for a typical day in the month of April is shown in Figure 5.2.

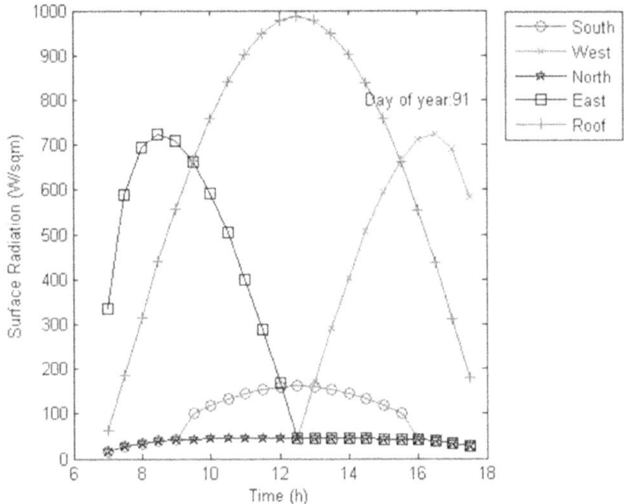

Figure 5.2: Radiation flux on walls and roof for a day in the month of April.

Following were inferred from Figure 5.2:

- The roof received higher amount of solar radiation flux (W/m^2) compared to other surfaces of the building envelope.

- Radiation on North and South wall was negligible.

- Radiation on East and West wall was symmetrical.

5.1.2 Building Length (L) / Width (W)

The effect of L/W ratio on the incident annual radiation on the building envelope, was studied. The L/W ratio was varied from 0.5 to 2 with the South and East facing walls taken as length and width respectively. The calculated annual solar radiation on the surface of the building envelope for different L/W ratios is as shown in Figure 5.3.

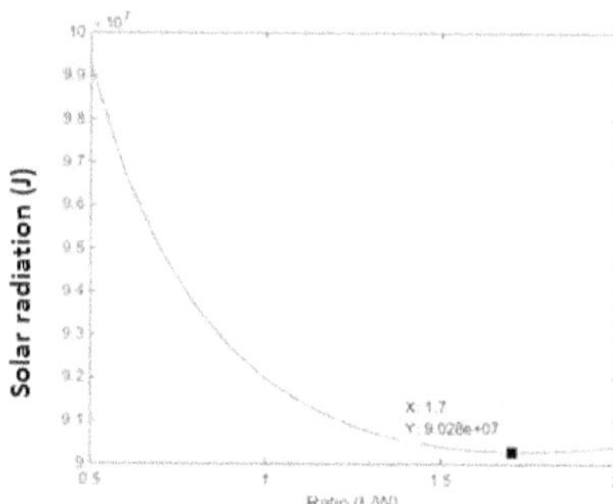

Figure 5.3: Annual surface radiation for different L/W ratios.

From Figure 5.3, following were observed:

- Buildings were generally constructed with a L/W ratio of 0.6. The surface radiation can be reduced by 9.4 % if an L/W ratio of 1.6 to 1.75 is used.

- It was found that an L/W ratio of approximately 1.618 (Golden ratio)[83], resulted in least annual surface radiation.

Based on the above mentioned inferences, an L/W ratio of 1.618 was used.

5.1.3 Building Orientation

The effect of building orientation on incident annual solar radiation was then studied. The building orientation was varied from 0^0 (longer side facing South) to 90^0 (longer side facing East). The results are as shown in Figure 5.4.

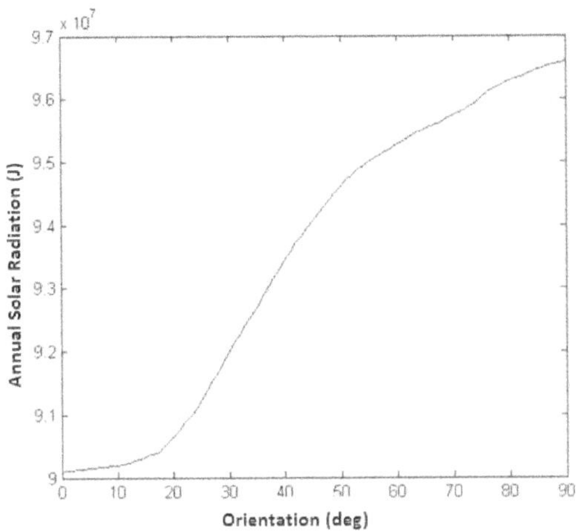

Figure 5.4: Annual Solar radiation for different building orientations.

Following were inferred from Figure 5.4.

- Zero degree orientation resulted in least amount of annual solar radiation.

- Radiation increased by 7.7 % for 90^0 when compared with that of 0^0 orientation.

5.1.4 Building Materials

Heat flow into the building was modulated using three thermal building material properties: a) Decrement factor (μ), b) Phase lag (Φ) and c) Transmittance (U).

Different combinations of building materials, given in the Table 5.1 were used for estimating the heat transfer (W) for a typical day in the month of April (hottest month in a year).

Type	Material	Time- Lag(hours)	Decrement Factor	U (W/m²-K)
Wall (W1)	Brick Plaster (130mm)	3	0.7	2.64
Wall (W2)	Brick Plaster (250mm)	7.6	0.33	1.83
Wall (W3)	Rammed Earth(300mm)	10.3	0.22	2.86
Roof (R1)	Concrete Roof (150mm)	7	0.58	0.89
Roof (R2)	Metal Corrugated Sheet	0.1	1	7.14

Table 5.1: Thermal properties of building materials

To calculate the periodic heat flow (Q), following equations were used[21].

$$T_{sol} = T_{hour} + \left(\frac{0.6S_T}{22}\right)$$

Eqn. 5.4

$$\Delta T_{net} = T_{avg} - T_{comfort} + \mu(T_{sol} - T_{avg})$$

Eqn. 5.5

$$Q = U.A.\Delta T_{net}$$

Eqn. 5.6

Where

S_T — average radiation on the surface [W/m²]

The daily energy requirement (Wh) for different combinations of building materials were calculated and is given in the Table 5.2. It was observed that Case 4 required minimum energy for cooling.

Case	Material	Energy requirement (Wh)
1	W2+R1	1505
2	W1+R1	5681
3	W3+R1	589
4	W3(white) + R1(white)	329
5	W2+R2	15660

Table 5.2: Energy requirements for different building material combinations

Since conduction through the roof contributed most to the heat transfer (Figure 5.2), a hypothetical roof with different decrement factors, Transmittance and Time lags were studied. For different thermal properties, energy requirement was calculated and is listed below in Table 5.3.

Following were observed:

- As decrement factor increased by 50 %, the cooling load increased by 29%.

- As thermal conductivity increased by 50%, the cooling load increased by 23%.

- As time lag increased by 38%, the cooling load decreased by 54%.

Increase in time lag resulted in substantial saving of energy, compared to properties like decrement factor and thermal conductivity.

Time Lag (h)	Decrement factor	U (W/m²-K)	Energy requirement (Wh)
7	**0.46**	0.89	288
7	**0.58**	0.89	329
7	**0.69**	0.89	372
7	0.58	**0.71**	294
7	0.58	**0.89**	329
7	0.58	**1.07**	364
4	0.58	0.89	411
7	0.58	0.89	329
10	0.58	0.89	188

Table 5.3: Energy requirement for a hypothetical roof

5.2 Earth Air Heat Exchanger (EAHE)

EAHE system is a pipe buried beneath the earth surface at a depth ranging from 1.5m to 3.0m. At such depths the soil temperature is approximately equal to that of the mean of annual temperature for that geographical location[45]. Ambient air is circulated through the pipe, using blowers. Whenever the temperature of the circulated air is more than that of the soil temperature, heat transfer takes place by convection. Under such conditions, the pipe outlet temperature will be less than that of the inlet and can be used for cooling buildings. The basic principle of EAHE is as shown below.

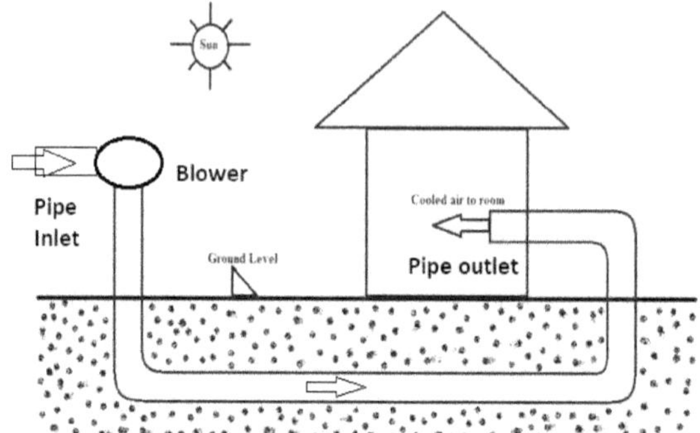

Figure 5.5: Basic principle of EAHE

The decrease in pipe exit temperature depends on the following parameters:

- Soil temperature at the given depth.

- Thermal diffusivity of the soil.

- Thermal conductivity and dimensions of the pipe (diameter, length).

- Air velocity.

Soil Temperature at a given depth and time of the year is given by

$$T_s(t,z) = T_a + Ae^{-z/d}\cos\left(wt - \frac{z}{d}\right)$$

Eqn. 5.7 [23]

Where

T_a –   Ambient temperature [°C]

z –   Depth [m]

d –   damping factor $= \sqrt{\left(\frac{2\alpha}{w}\right)}$

α –   diffusivity of soil $\approx 0.139 \times 10^{-6}\,m^2/s$

w –   Angular frequency variation $= 2\pi/\,(24 \times 60 \times 60) =$
$0.727 \times 10^{-4}(s^{-1})$

A –   Amplitude of ambient temperature variation [°C]

From above equation. it is seen that diurnal variations are damped out. as the depth increases. The yearly variation of ground temperature for Coimbatore is as shown in Figure 5.6 below. It is observed from Figure 5.6 that for April, the soil temperature is 25°C which was lesser than the upper limit of comfort temperature of 31°C. Hence EAHE can be used as a natural cooling method in Coimbatore.

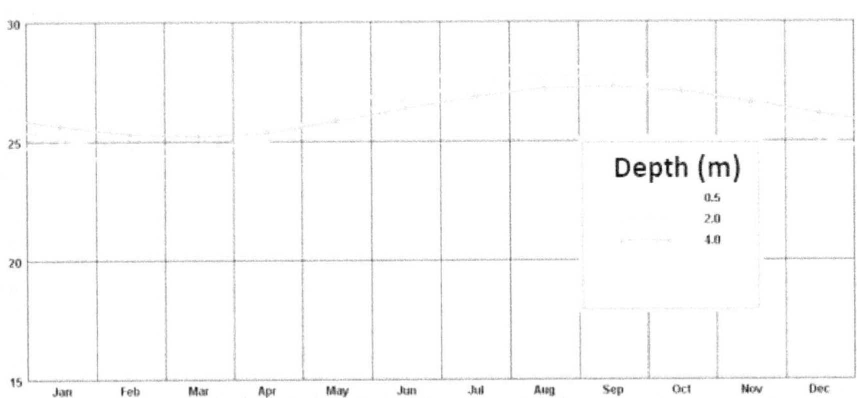

Figure 5.6: Annual variation of ground temperature at Coimbatore[84]

5.2.1 EAHE modelling

For a circular pipe of diameter (D) and length (L) following equations[85] were used to estimate the pipe outlet temperature (T_{outlet}), for a given soil temperature (T_{soil}).

$$T_{outlet} = T_{soil} + (T_{inlet} - T_{soil})e^{-NTU}$$

Eqn. 5.8

Where Number of Transfer Units (NTU) for a given heat transfer area (A) and mass flow rate ($\dot{m}$) is given by the relationship

$$NTU = \frac{hA}{\dot{m}C_{air}}$$

Eqn. 5.9

Convective heat transfer coefficient (h) for the pipe material conductivity (K) is calculated from the Nusselt number (Nu)

$$Nu = \frac{hD}{K}$$

Eqn. 5.10

Nu for a given Reynolds number (Re) and Prandtl number (Pr) is given by the following equation

$$Nu = 0.023.R_e^{0.8}.(Pr)^{\frac{1}{3}}$$

Eqn. 5.11

The power required (W) to ensure a constant flow rate (v) is given by the following

$$\dot{W} = v\Delta P_L$$

Eqn. 5.12

Where ΔP_L is the pressure drop along the pipe given by the following equation

$$\Delta P_L = f\left(\frac{L}{D}\right)\left(\frac{1}{2}\rho V_{avg}^2\right)$$

Eqn. 5.13

f is the Darcy factor given by $f = 0.184R_e^{-0.2}$,

For the circular pipe of cross section area (A_c) and mass flow rate of air ($\dot{m}$) , average velocity is given by following equation

$$v_{avg} = \frac{\dot{m}}{\rho A_c}$$

Eqn. 5.14

5.2.2 Experiments with EAHE:

Study of Sub-Terranean Temperature

An experiment was carried out in Coimbatore. A trial pit was made with a dimension 1.2m(L) by 1.2m(W) by 2.13m(H). A PVC pipe of 2.13m (7ft) was taken and two holes were drilled at 0.3m (1ft) and 1.2m (4ft) length. Three 'BT18D20' temperature sensors were placed in the pipe at 1ft, 4ft and 7ft depth. The remaining region inside the pipe was covered using cotton restricting the passage of airflow through the pipe. The pipe was then buried vertically inside the pit. The trial pit is as shown in the Figure 5.7 below.

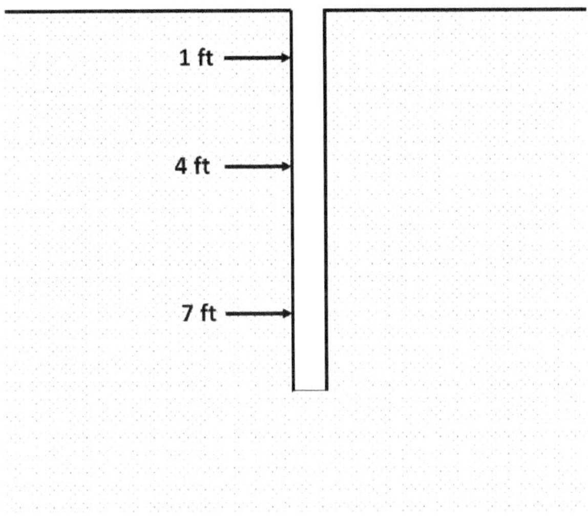

Figure 5.7: Trial pit with three temperature sensors

Temperatures inside the trial pit were recorded for three months. The variation observed for the month of February 2016 is as shown in Figure 5.8.

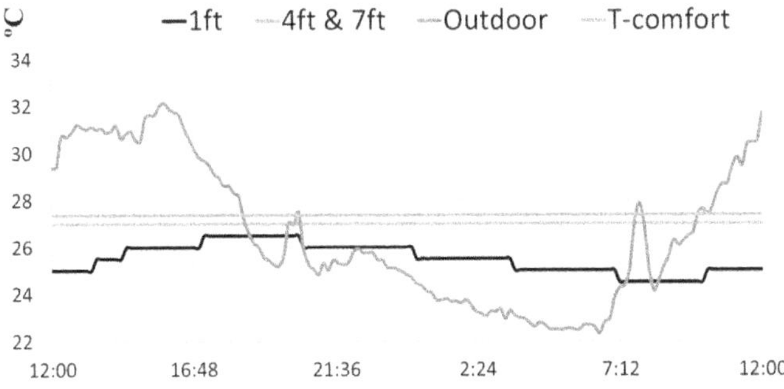

Figure 5.8: Diurnal temperature variations (February 2016) at different depths

The amplitude of the temperature diurnal variation reduced with increase in depth and was negligible at 7ft. It was also observed that both 4ft and 7ft temperature variations were similar throughout the day. They were also close to the thermal comfort level for February. Hence it was felt that the site was suitable for installing the EAHE.

EAHE Model – I

An EAHE was then installed at the same site. It was an open loop EAHE system which used a 0.5hp centrifugal blower, operating at minimum outlet velocity. A PVC pipe of 12m (40ft) length) and 0.23m (9 inches) diameter was laid at a depth of 1.21m (4ft).

OBSERVATIONS:

- The outlet temperature of the pipe was 4.5°C lesser than that of outside temperature on an average during the month of March. This value was in agreement with the theoretical model.

- As summer approached, the temperature at 4ft increased rapidly. The temperature was 31°C by the end of April. The mean temperature variation at 4ft depth for the months of January, February, March and April is shown in the Figure 5.9 below. Since the soil temperature was above the thermal comfort level, the experiment was discontinued.

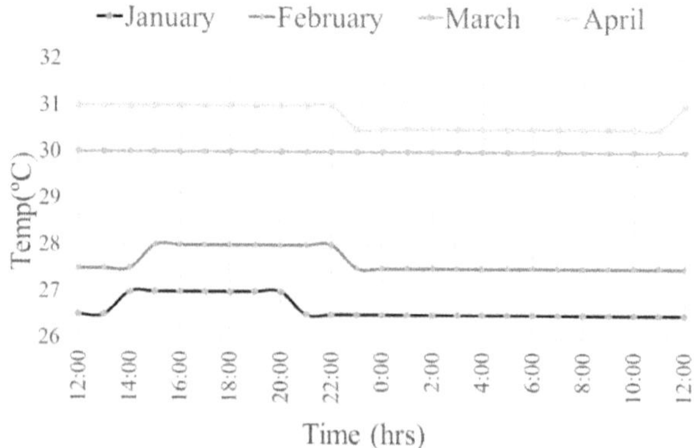

Figure 5.9: Mean temperature variation at 4ft depth from January to April

As EAHE-I did not provide satisfactory results, the next alternative was to lay the pipes deeper. But this would have increased the cost of installation.

EAHE Model – II

The system was redesigned with a pipe of 24m (80ft) length and 0.1m (4 inches) diameter, buried in a trench of 0.45m (1.5ft) depth. The centrifugal pump which was used in the previous model was used with a minimum velocity of 3.5m/s.

Soil conditioning[SC] was used over the trench, with a 0.1m layer of gravel. It was watered for a period of 20 minutes in the morning daily to increase the moisture content. The EAHE outlet temperature and outdoor temperature recorded during the experiment (May 2016) is as shown in Figure 5.10.

It was found that the pipe outlet temperature was constant at 29.5°C and always lower than the comfort temperature. The maximum difference between outdoor and EAHE outlet temperature was found to be 5.5°C. **The experiment indicated that EAHE with soil conditioning is a viable natural cooling option.**

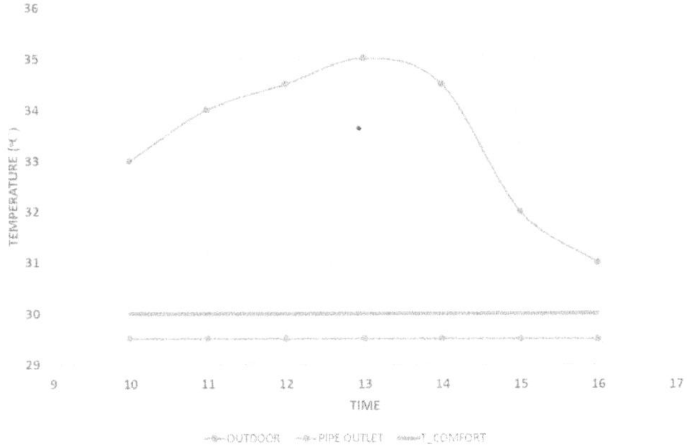

Figure 5.10: Performance of EAHE during summer (May 2016).

Chapter 6

Natural Ventilation

Natural ventilation significantly reduced the indoor cooling load, if the outdoor temperature is less than that of the upper limit of the comfort temperature[28].

Natural ventilation depended on the pressure difference between openings in the building envelope. The pressure difference was due to: a) Outside wind b) Buoyancy driven flows inside the building. The air flow rate (Q) for outside wind and buoyancy driven natural ventilation is given in Figure 6.1[30].

Where,

A1, A2, A3 and A4	–	Area of openings.
H	–	Height difference between openings.
ΔC_p	–	Difference in pressure coefficient between windward and leeward sides.
C_d	–	Discharge coefficient.
u_r	–	Outdoor wind velocity.

Conditions	Schematic representation	Formula
(a) Wind only		$Q_w = C_d A_w u_r (\Delta C_p)^{\frac{1}{2}}$ $\dfrac{1}{A_w{}^2} = \dfrac{1}{(A_1 + A_2)^2} + \dfrac{1}{(A_3 + A_4)^2}$
(b) Temperature difference only		$Q_b = C_d A_b \left(\dfrac{2\Delta\theta g H_1}{\bar{\theta}} \right)^{\frac{1}{2}}$ $\dfrac{1}{A_b{}^2} = \dfrac{1}{(A_1 + A_3)^2} + \dfrac{1}{(A_2 + A_4)^2}$

Figure 6.1: Air flow rates for wind only and buoyancy driven natural ventilation[30]

It was evident from above table, that air flow rate due to buoyancy effects was a strong function of the height difference between two openings. In present generation buildings, the floor to ceiling height has been generally less than 10ft and buoyancy driven flow rates were negligible. **Hence, only wind driven natural ventilation was considered.**

6.1   Wind driven natural ventilation

Santamouris[23] studied the effect of winds around buildings on air flow inside the building. He observed following:

a) Positive and negative pressure regions developed on the windward and side walls as shown in Figure 6.2 below

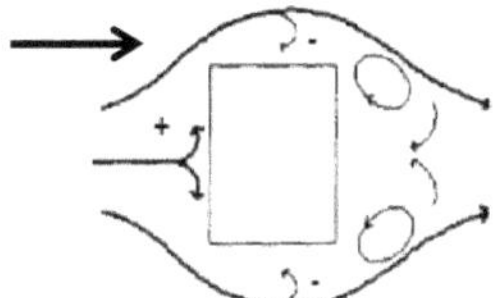

Figure 6.2: Airflow patterns around a building without windows[23]

b) A jet developed if openings [windows] are placed on both the windward and leeward walls, as shown in Figure 6.3 below.

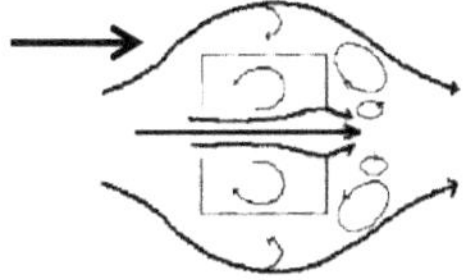

Figure 6.3: Ventilation through windward and leeward openings[23]

c) Cross ventilation improved if the openings were on the adjacent walls as shown in Figure 6.4. Internal circulation was also better compared to previous option.

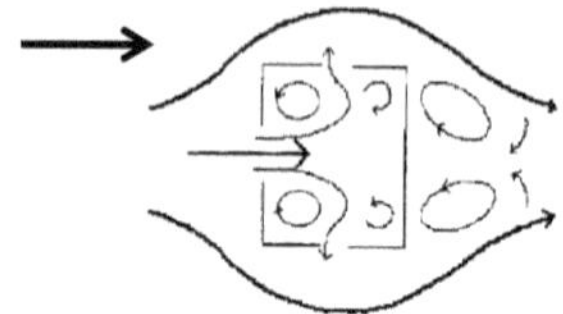

Figure 6.4: Ventilation through windward and side openings[23]

Pressure coefficient (C_p): Airflow inside the room was a strong function of the pressure coefficients of the windward and leeward sides as shown in Figure 6.1. However estimation of C_P using analytical equations has been difficult because it depended on the following[36].

- Building geometry.

- Balconies, sunshades.

- Window position on the building external surface.

- Neighbouring buildings and trees.

- Wind speed and direction.

- Turbulence intensity.

Wind tunnel experiments on scaled down models of buildings, were often used to predict C_p values. Twelve different institutions carried out experiments on a simple cube using wind tunnels at different locations[87]. Wide variations were reported in the results as shown in   Figure 6.5 below.

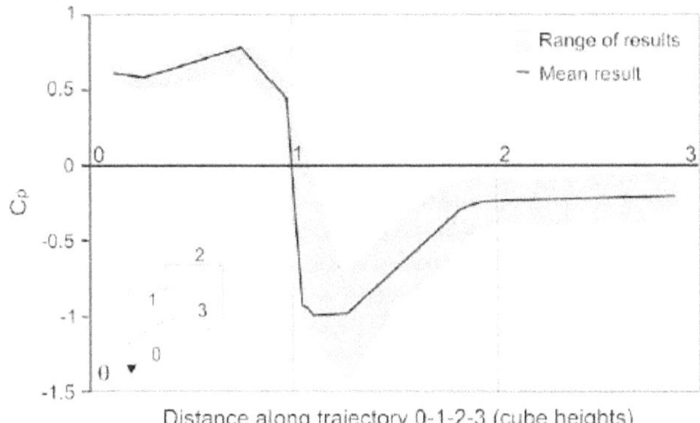

Figure 6.5: Variation of wind tunnel data recorded at different wind tunnels[87]

Due to difficulties in development of analytical models and uncertainties in wind tunnel experiments even for simple shapes like cubes, CFD was used for predicting C_P values[88]. Moreover CFD also gave detailed information on the internal airflow. CFD was used also for analysis of effect of complicated geometrical structures called wing walls on internal airflow.

6.2   Work carried out on Wing walls (TVK Sushil Kumar)[40]

Wing walls are external projections on building external surfaces adjacent to windows. When positioned properly, the difference between the pressure coefficients at the inlet and outlet windows, (ΔC_P) increased. This improved cross ventilation even during low external wind velocity conditions. Dekay[39] gave subjective ratings (scale from best to poor) on the direct flow pattern between the inlet and outlet windows, with wing walls.

Graca[89] proposed empirical formulae for a) Estimating the air velocity in a room (between the inlet and outlet), b) Velocity in the recirculation regions.   Mak[90] carried out a CFD study on effect of wing walls on indoor air velocity.

Both researchers studied effect of windows placed on opposite walls. Windows are usually placed on adjacent walls for houses having multiple rooms for better internal air circulation as shown in Figure 6.3. Natural ventilation inside a house with wing walls and windows placed on adjacent sides was studied.

6.2.1 Validation

Results given by Graca[89] were used to validate the CFD model. A single room house with a square plan form of 4.5m and a height of 3.5m, used by him, was simulated. Both the inlet and outlet windows were placed at the centre of geometrically opposite walls and were of $1m^2$ area each.

Graca has not given details about the external volume used in his simulations. However, a suitable external volume should be defined. Hence the room mentioned above, was enclosed by an external volume, according to best practices recommended, for architectural applications. The CFD model used for the study is as shown in Figure 6.6.

Figure 6.6: CFD model used for simulation (all dimensions are in SI units).

Wind velocity (5m/s) and zero-gauge pressure were used as inlet and outlet boundary conditions for the external volume. A free space environment was defined by assigning slip condition for top and sides of the external volume. Adaptive meshing was used to ensure finer mesh in high gradient regions and coarser meshes in rest of the regions. Comparison between velocities obtained using Graca formulae and simulations are listed as Table 6.1.

Results of simulation was found to be in close agreement Graca formulae

Inlet velocity (m/s)	Main jet velocity (m/s)		Recirculation velocity (m/s)	
	Graca	Simulation	Graca	Simulation
1.0	0.78	0.80	0.17	0.17
1.5	1.17	1.10	0.26	0.26
2.0	1.56	1.50	0.34	0.34
2.5	1.96	1.90	0.43	0.40

Table 6.1: Comparison of velocities

6.2.2 Effect of Wing Walls on Natural Ventilation

Square windows of side 1.22m (4ft) were placed on adjacent walls of a room of length 4.87m and width of 3.33 m. The winged walls were of 1.22 m in length and 4.27 m in height.

Four different configurations were studied. The results are as shown in Figure 6.7. Dark shades in Figure 6.7 indicate low velocity regions.

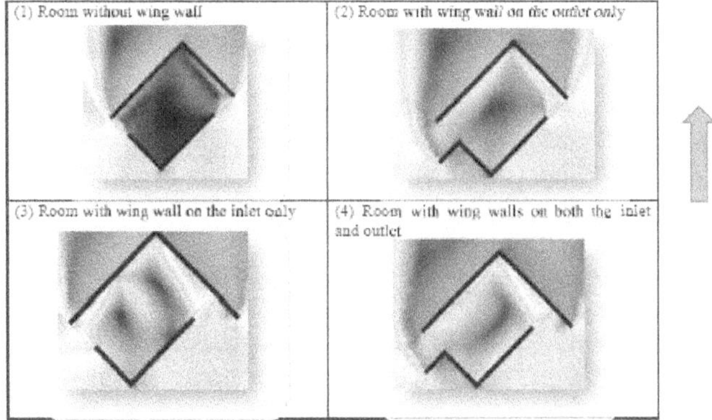

Figure 6.7: Effect of Wing walls on natural ventilation for different configurations.

Average inlet velocities for different configurations are given in Table 6.2 below.

It was observed that a single wing wall on the inlet side (third configuration) can result in 92% of the velocity achievable with two wing walls.

Configuration	Average inlet velocity (m/s)
1	2.64
2	4.1
3	**4.7**
4	5.12

Table 6.2: Average inlet velocity for different wing wall configurations

6.2.3 Study of Air Flow Pattern for Different Wing Wall Positions:

Rooms with an inlet and outlet on adjacent walls experienced two regions of flow namely a wall jet (a jet which flows along the surface of a wall) and regions of recirculation. Velocity in the wall jet region was much higher compared to the recirculation region, where the flow was almost stagnant. Such low velocity regions can result in thermal discomfort.

For different wing wall configurations, Dekay[39] has only given guidelines for direct streamlines between the inlet and outlet. Besides obtaining information on the flow between the inlet and outlet, the simulations have also identified recirculation regions. These regions were characterized by very low velocities and are shaded black in Figure 6.8 below.

Average inlet velocities obtained for eight configurations shown in Figure 6.8 below is as listed in Table 6.3.

Configuration	Mark Dekay rating	CFD results (inlet velocities) m/s
1	Best	3.96
2	Good	3.0
3	Poor	0
4	Poor	1.0
5	Best	3.78
6	Good	3.3
7	Poor	0
8	Poor	1.9

Table 6.3: Comparison between Mark Dekay rating and CFD results

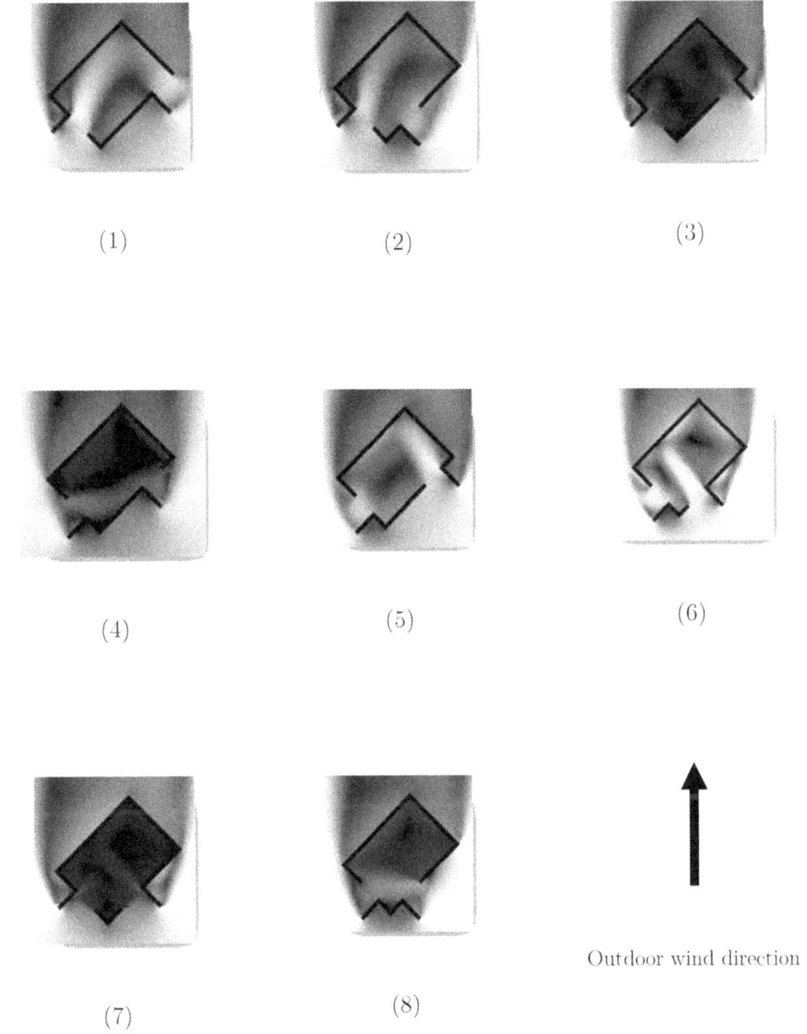

Figure 6.8: Flow visualization for different Wing wall positions.

Results obtained through CFD simulations were in agreement with the ratings given by Dekay.

6.2.4 Effect of Wing Length on Average Inlet Velocities

For a given position of inlet and outlet as shown in Figure.6.9, wing wall length 'L' on inlet velocity

was studied.

Figure 6.9: Effect of wing wall length (L) on flow pattern.

Inlet velocities with increasing wing wall lengths and its construction cost is as shown in 6.10.

Construction cost was taken as Rs. 3225-/- for every square meter. It was observed that, a

0.45 m wing wall length, was sufficient to achieve 90 % of the velocity obtained using a wing wall

length of 1.8 m, at 25 % cost.

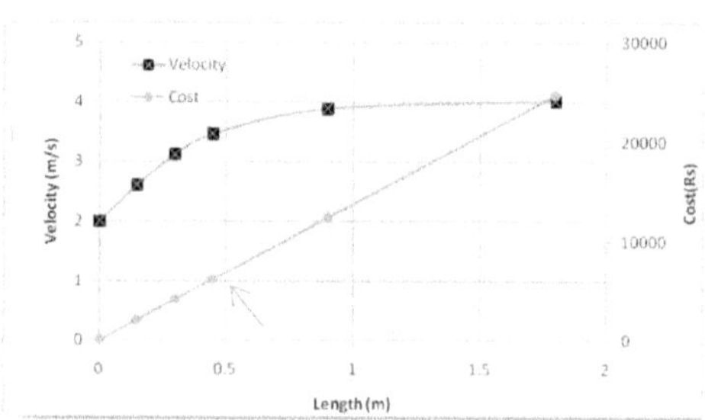

Figure 6.10: Effect of wing wall length on inlet velocity and construction cost.

6.2.5 Wing Wall Height:

Above mentioned simulations were carried out using a wing wall of height 4.27 m, which is the

height from the ground level to the top of the parapet wall. Construction of such high walls was

not considered economical. Hence, the effect of wing wall height on inlet velocity, was studied. In this study the wing wall length was maintained constant at 0.45 m. Results are as shown in Figure.

6.11.

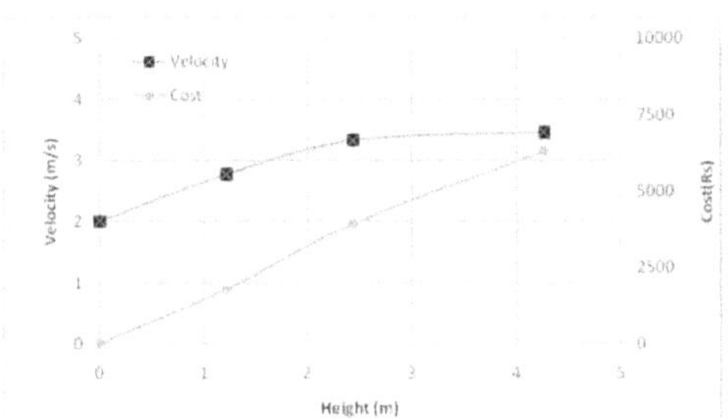

Figure 6.11: Effect of wing wall height on inlet velocity and construction cost.

It was observed that a wing wall of same height as that of the window (1.22 m) resulted in 80 % of the inlet velocity obtained using a full height wing wall (4.27 m), at 28% cost.

6.2.6 Inverted 'L' Type Wing Wall

An inverted 'L' type wing wall not only improved the aesthetics but also provided shading. Hence a study was carried out on an inverted "L" wing wall of dimensions shown in Figure 6.12.

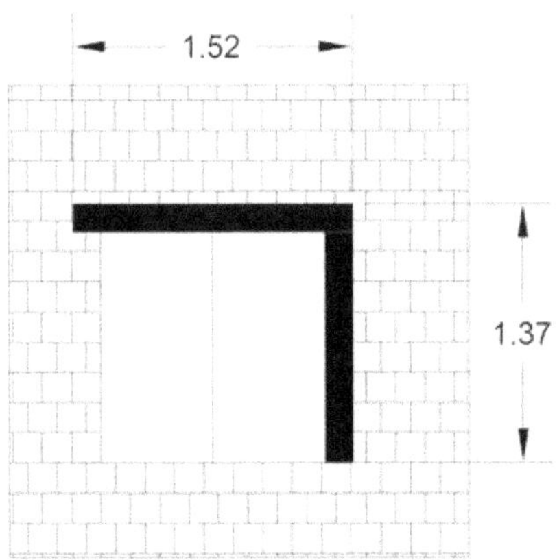

Figure 6.12: Front view of Inverted 'L' type of wing wall

An average inlet velocity of 3.17 m/s was observed, which is 90% of the velocity of a full height wing wall.

Observations:

- An inverted 'L' type wing wall shown in Figure 6.12, was a better option compared to a full height wing wall.

- It was sufficient to improve thermal comfort, under outdoor wind velocity of 5m/s.

- The inverted 'L' type wing wall can be realised by opening the appropriate window by the occupants depending on the prevailing wind direction. To prevent leakages, blanking can be fitted between the window outer frame and the hinged window edge. Hence it will require negligible additional cost.

Chapter 7

Nocturnal Cooling

Nocturnal cooling is a natural ventilation method by which the heat gained inside the building is released during nights with the help of cooler outdoor air. It is also called as night time cooling. The cool outdoor air replaces the hot air inside the building and also cools the internal surfaces of the walls and ceiling.

Nocturnal cooling reduced the rate of temperature increase during daytime[91][92] . This in turn reduced the energy required for cooling and improved the thermal comfort. Important parameters that affect nocturnal cooling are air flow, thermal mass, temperature differential and Convective Heat Transfer Coefficient (CHTC).

7.1.1 Air Flow

Velocity of airflow determines the Air Change per Hour (ACH) and gives the volume of air that is replaced every hour. It is calculated using the following formula.

$$\text{ACH} = \frac{v_{air} \times A_{outlet} \times 3600}{\forall_{room}} \quad [hours]$$

Eqn. 7.1

Where

v_{air} — Air velocity at the outlet

A_{outlet} — Outlet area

$\forall_{room}$ — Volume of the room

Since the outdoor air velocity is highly unpredictable, an ACH range 0.2 - 4 was observed in naturally ventilated buildings[93]. It has also been reported[94], that if the requirement for ACH is

greater than 15, nocturnal cooling is not a viable option. The cooling potential of nocturnal cooling
is calculated using following equation.

$$\theta_{NV} = \int_{t_i}^{t_f} \dot{m}_{air} C_{air} (T_{outdoor} - T_{exhaust}) dt$$

Eqn. 7.2

Where

$\dot{m}_{air}$ —   mass flow rate of air

$T_{exhaust}$ —   Temperature at the outlet of the exhaust fan

7.1.2 Temperature differential and Thermal mass:

It was observed from equation 7.2 above, that nocturnal cooling is a strong function of the
temperature differential, for a given mass flow rate. Shaviv[95], carried out experiments to ascertain
the potential of nocturnal cooling. It was observed that reduction of maximum indoor temperature
(T_R) was a function of diurnal temperature variation and thermal mass. His results are summarised
in Figure 7.1 below.

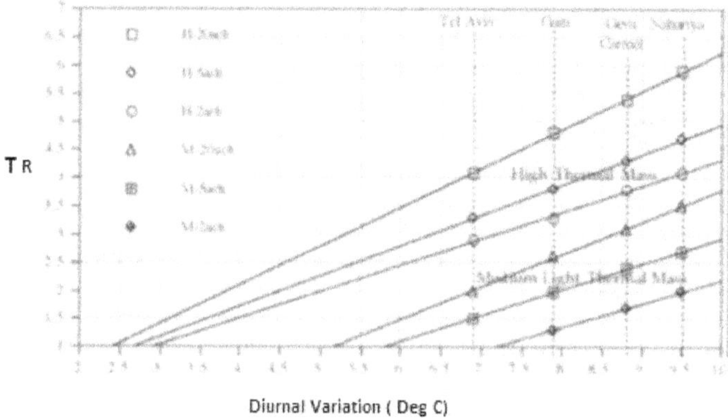

Figure 7.1: Effect of diurnal variation and thermal mass of building on indoor temperature[95]

Shaviv reported that a high thermal mass building reduced the maximum indoor temperature considerably. The reduction in indoor maximum temperature of a high mass building was 50% more than that of a low mass building. Further, the peak indoor temperature was found to be a strong function of the outdoor average temperature and not the outdoor peak temperature.

7.1.3 Convective Heat Transfer Coefficient (CHTC):

During night-time the internal surface temperature of the walls and ceiling can be decreased by effective interaction between the cool outdoor air and the surface. Heat transfer by convection depends on the CHTC.

It has been reported[96] that small volumes promoted a homogenous airflow distribution inside the rooms and raises the average convective heat transfer coefficients. This in turn improved the night-time cooling of internal surfaces.

7.2 Barriers to night ventilation:

It was observed that in a naturally ventilated building at Coimbatore, the outdoor temperature is less than the indoor temperature from 1800h in the evening to 0800h next day morning. In spite of the cooling potential of natural ventilation during this period of, several barriers have been reported[43].

- Occupants are reluctant to open windows during night-time.

- Intrusion of insects and snakes.

- Fear of robbery

- Ingress of rainwater

- Sand storms in desert areas

- Pollutions in urban areas

- Noise in buildings adjacent to busy roads and markets

- Outdoor temperature is not substantially lower than indoor temperature for sufficient duration in some geographical locations, during summers.

7.3 Work carried out on ascertaining the nocturnal ventilation for Coimbatore.

Out of the barriers for nocturnal ventilation mentioned above, the last one was analysed for Coimbatore, using Energy Plus Weather (EPW) database and 'Climate consultant' software. Monthly and diurnal variation of outdoor temperature, average wind speed and direction are shown in Figure 7.2 and 7.3 below.

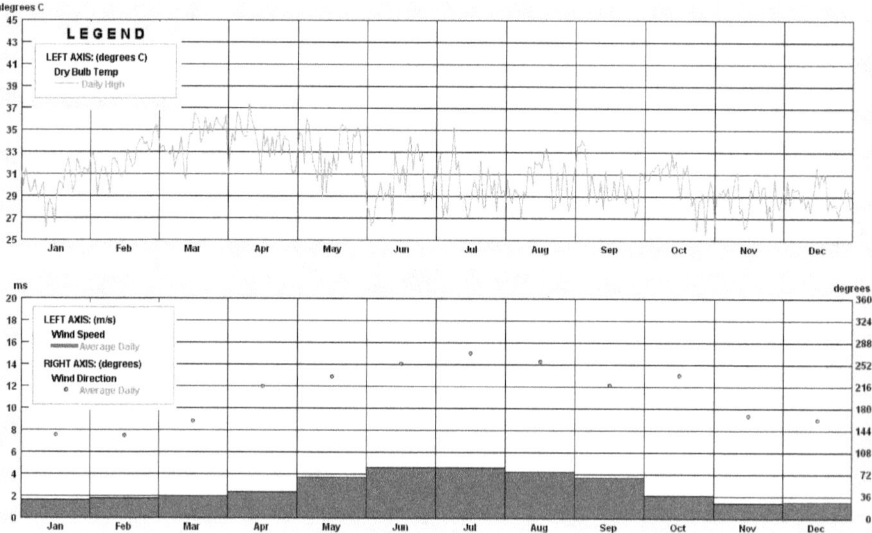

Figure 7.2: Monthly variation of maximum outdoor temperature, wind velocity and direction

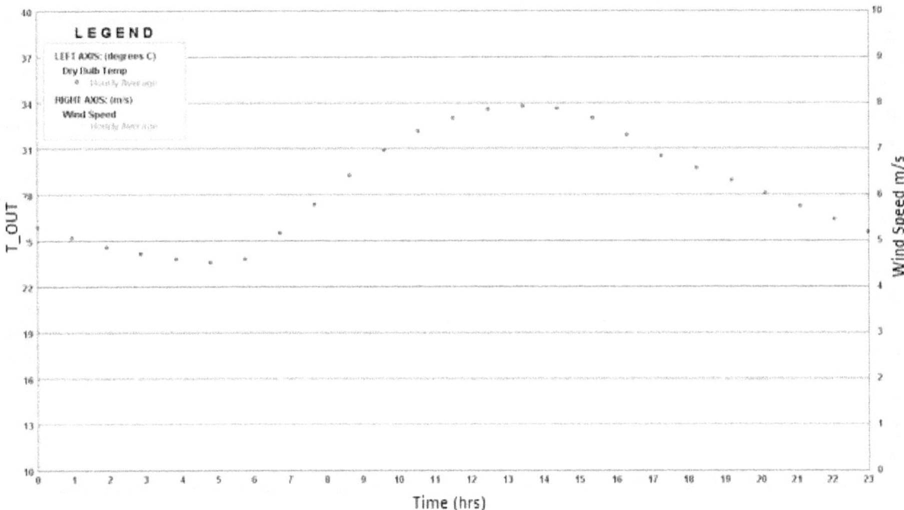

Figure 7.3: Diurnal variation of average outdoor temperature and wind velocity in April.

Following were observed from Figures 7.2 and 7.3.

- For the summer month of April, the wind velocity was 2m/s compared to 4.25m/s during the cooler months of June and July (Figure 7.2).

- Though the outdoor temperature was lower than the indoor temperature from 1800h to 2100h, the wind velocity reduced to 2m/s, thereby reducing the natural ventilation potential of the building (Figure 7.3).

70

Chapter 8

Controlled Nocturnal Cooling

Nocturnal cooling had several barriers. Manual intervention by the occupants was required to overcome these barriers: a) Indoor and outdoor temperatures had to be monitored b) Appropriate windows had to be opened whenever $T_{out} < T_{in}$. Given the present-day busy lifestyle and high expectations, there has been very less evidence of such initiatives having been taken by the occupants.

Hence there has been a need to automate the whole process. Since early nineties, researchers like Givoni and Roche have tried to use controlled nocturnal cooling, to improve thermal comfort levels.

Givoni[58] carried out experiments on two rooms each having a floor area of $23m^2$(250 sqft). The thermal mass of the rooms were different. One used stud wall (low thermal mass) and the other solid concrete (high thermal mass). Windows were opened manually from 7pm to 7am and a controlled night ventilation was achieved using exhaust fans running at three different speeds. This enabled variation of Air Change per Hour (ACH): 30ACH for low speed, 37ACH for mid speed and 45ACH for high speed. He observed reduction in indoor night-time temperatures for both low and high thermal mass rooms, which also resulted in reduction of average indoor temperature for the day. He observed that indoor peak temperature reduced significantly only for the high mass building. Based on his experiments, he formulated following empirical relation to estimate maximum indoor temperature (T_{max}), for naturally ventilated buildings.

$$T_{max} = GT_{avg} + DelT + k(T_{avg} - GT_{avg})$$

Eqn. 8.1

Where values of **DelT** and k depended on building thermal mass and use of natural ventilation. GT_{avg} is the average outdoor temperature for the month and T_{avg} is the mean outdoor temperature for a particular day.

Equation 8.1 has been used to estimate the maximum indoor temperature at Coimbatore and compared with experimental results (Table 8.1). It was observed that Givoni formula overestimated the indoor temperature.

	T_{avg}	Estimated	Measured
Day 1	28.12	30.89	29.69
Day 2	29.52	31.59	29.87
Day 3	30.7	32.18	30.69

Table 8.1: Comparison between estimated and measured indoor maximum temperatures

Roche[44] carried out experiments on two test cells of identical dimension (4ft wide, 8ft long and 8ft high). The 'control test cell' did not have any provision for ventilation. The 'experimental test cell' on the other hand was ventilated with the help of an exhaust fan. The exhaust fan was capable of producing 3.9 to 15 ACH. It was controlled based on the following logic.

If $T_o < T_{in}$ AND $T_{in} > T_{comfort-High}$, then activate the exhaust fan. The upper limit of comfort temperature ($T_{comfort-High}$) was set at 25.5 °C. Results of his experiment is as shown below in Fig 8.1.

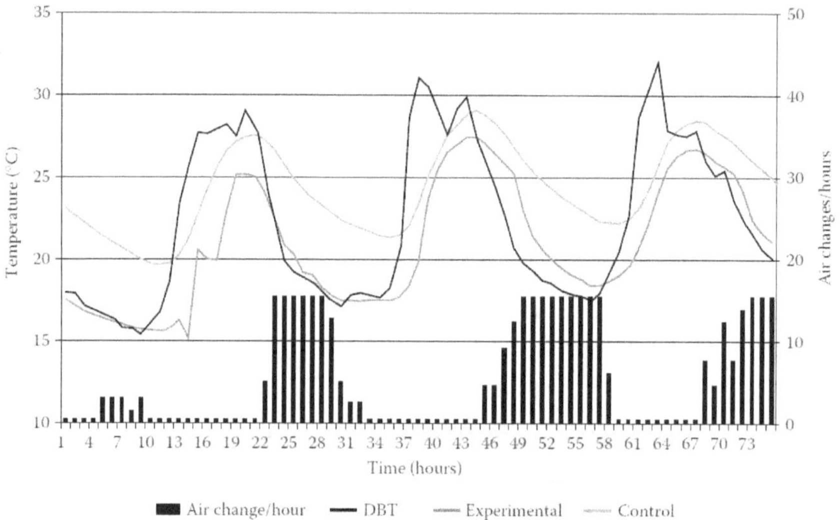

Figure 8.1: Effect of controlled nocturnal cooling on indoor temperature[14]

Following were observed from his experiments.

- The upper limit for comfort temperature was fixed at a constant value of 25.5°C irrespective of the geographical location and day of the year. Hence it was not in accordance with the adaptive comfort temperature model.

- Indoor temperature exceeded **upper limit of comfort temperature** for nearly 8 hours in a day (1600h to 2400h).

- Based on simulations, he also reported that his model was not effective in four out of sixteen climatic zones of California.

Roche model reduced energy consumption when compared with conventional air-conditioner with thermostat control [fixed setting of indoor temperature according to Fanger model]. It however did not prevent over heated periods in all climatic zones. Hence, there was a requirement to use a combination of controlled nocturnal cooling and EAHE, based on upper comfort temperature limits

recommended by the adaptive comfort model.

Chapter 9

Proposed analytical and ANN based building heat transfer model

Limitations of building-heat-transfer models discussed earlier in Literature review were due to the following:

- Heat conduction across the building envelope was calculated using the equivalent external surface temperature '**Sol-air temperature**', which was difficult to measure.

- Periodic heat flow and one dimensional models were based on availability of data only on building material properties like decrement factor and time lag.

- The internal temperature (Fanger model) was held constant and did not change over a period of time i.e. occupants were not willing to adapt to changing climatic conditions (geographical, diurnal and annual).

Due to the limitations mentioned above, it was felt that existing models were useful for estimating cooling load only in a controlled environment, which used conventional air conditioners and not for passively cooled buildings in which the occupants were willing to adapt themselves to changing climatic conditions.

9.1 Sol-Air temperature

Simulations were carried to find the average diurnal difference between T_{Solair} and T_o, on all external surfaces, for a house at Coimbatore, with plastered wall and light coloured roof. Results of the simulation are given in Table 9.1 below.

Walls				Roof	Average
North	East	South	West		
0.2°C	1.7°C	0.5°C	1.5°C	-0.1°C	0.8°C

Table 9.1: Average diurnal difference between Sol-air and outdoor temperatures

It was observed that, on an average, T_{Solair} for all external surfaces exceeded T_o only by 0.8°C, during summer. Hence, a constant upward shift of 2°C in T_o was assumed to take in to account worst case scenarios. This assumption was also subsequently annulled with the help of an ANN based model.

A new analytical model is proposed with emphasis on following.

- Inputs: Only date, time, outdoor and indoor temperatures. These are easily measurable/ logged parameters compared to Sol-air temperature.

- To reduce the indoor temperature, natural heat sinks like outdoor air (nocturnal cooling) and cooler soil temperature (EAHE) are utilised.

9.2 The Proposed model

This model is based on solving the following differential equation which allows for variation in indoor temperatures depending on external weather conditions.

$$\frac{dT_{in}}{dt} = \frac{\left(\frac{dQ}{dt}\right)_{gain} - \left(\frac{dQ}{dt}\right)_{loss}}{m_{air-room} \cdot C_{air}}$$

Eqn. 9.1

Where

$\left(\frac{dQ}{dt}\right)_{gain}$ — Rate of heat transfer in to the room due to conduction through the walls and roof.

$\left(\frac{dQ}{dt}\right)_{loss}$ — Rate of heat transfer out of the room with the help of natural heat sinks (EAHE or nocturnal cooling).

$m_{air-room}$ — Mass [kg] of the air inside the room which is the product of its density and volume.

C_{air} — Specific heat of air [J/(kg*K)].

The Simulink model for solving above differential equation is as shown in Figure 9.1. Based on measurements carried out at Coimbatore, the indoor and outdoor temperatures were found to be equal to the upper limit of comfortable temperature, at 1800h and the same was set as initial condition for the integrator. A constant occupant load of two persons generating 80W each was taken.

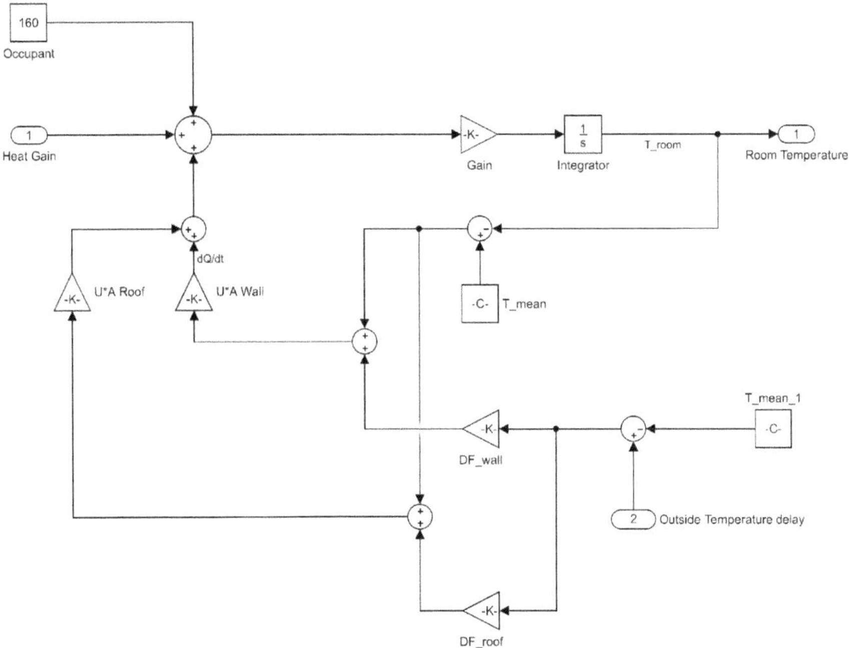

Figure 9.1: Simulink model for solving equation 9.1.

Where

U : Thermal transmittance of the walls and roof $[W/m^2\text{-}K]$

A : Area of the walls and roof $[m^2]$

K_1, K_2 : product of C_{air} and mass flow rate[kg/s] of cooler air from EAHE ground

and bypass mode respectively.

$K_3 = (m_{air\text{-}room} * C_{air})^{-1}$

Outside temperature is delayed due to time lag[hours]

DF: decrement factors of wall and roof.

9.3 Validation of the proposed analytical model

Simulations were carried out for a typical summer day in April and results are as shown in Figure
9.2 below. Outdoor temperatures for Coimbatore were obtained from EPW files.

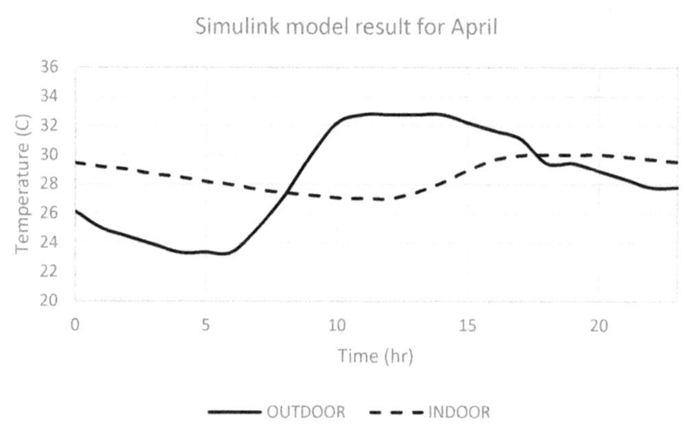

Figure 9.2: Simulated Indoor temperature for the month of April

Measurements were then carried out for the outdoor and indoor temperatures inside a room of

similar dimensions for 24hours, for the month of April. Variation of measured indoor temperature

with diurnal variation of outdoor temperature is as shown if Figure 9.3 below.

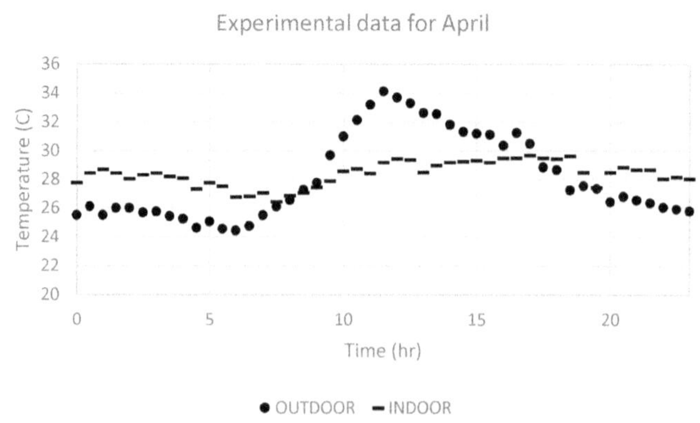

Figure 9.3: Measured Indoor and outdoor temperatures for the month of April

It was observed that the proposed analytical model results captured the decrement factor and time lag as shown in Table 9.2 below.

Parameter	Measured	Simulation
Time lag(hrs)	6	6.5
Decrement factor	0.88	0.91

Table 9.2: Comparison between Measured and simulated Time lag and decrement factor

9.4 Artificial Neural Network (ANN) based indoor temperature model

In order to reduce dependence on time lag and decrement factor values, an ANN was trained with date, time, outdoor and indoor temperatures. The house model was simulated as an ANN problem using MATLAB (Simulink), with easily measureable parameters like outdoor and indoor temperatures ,day of the year and hour of the day as inputs and the next hour indoor temperature (T_{in+1}) as the output. Measurement of $T_{Sol\,air}$ was not needed as the ANN model is a mapping of outdoor temperature and next hour indoor temperature, for a given day of the year. Annual climate data for Coimbatore, was obtained from EPW files. Training was carried out for all 12 months using monthly average hourly temperatures i.e. 288 datasets (24 hours X 12 months). Simulink model for training the network is as shown in Figure 9.4 below.

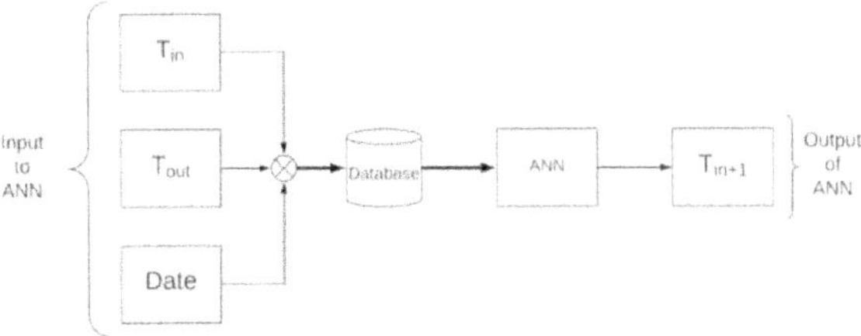

Figure 9.4: Simulink model for training the house model

9.4.1 Validation of the ANN based model

The measured outdoor temperatures were given as input to the ANN. The measured indoor temperature was compared with ANN predicted values and is as shown in Figure 9.5 below.

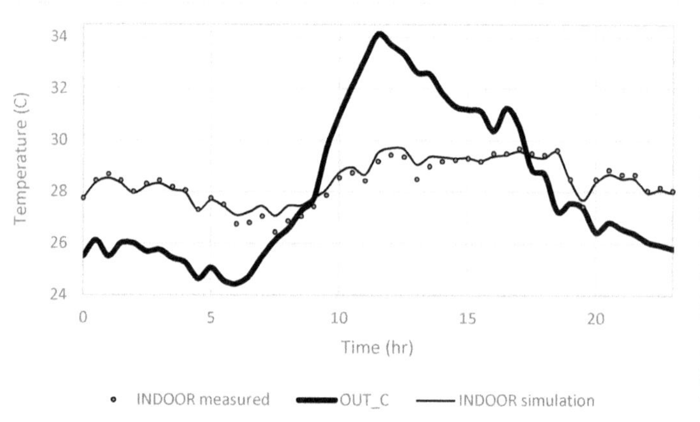

Figure 9.5: Comparison between measured and ANN predicted indoor temperatures

The predicted ANN indoor temperature was very close to the measured values with a RMS error of 0.18°C (standard deviation of 0.15). It clearly indicated that ANN is suitable for modelling and can be used for implementation of control logic.

Chapter 10

Control law to maintain indoor temperatures within adaptable comfort limits

As discussed earlier. an ANN model based on easily measurable parameters can be used for predicting the next hour indoor temperature. This ability of the ANN and availability of natural heat sinks (EAHE and nocturnal cooling) has been used to develop a control law. in order to maintain the indoor temperature within adaptive thermal comfort limits.

The adaptive indoor comfort temperature (T_C). recommended by Nicol[14] , was used for determining comfortable temperature limits. Unlike the ASHRAE model[15] which required estimation of MRT. Nichol defined comfort temperature based on the monthly mean outdoor temperature as given in the following empirical formula.

$$T_C = 17.0 + 0.38 * T_M \qquad \text{Eqn. 10.1}$$

Where

T_C –   Comfort temperature in °C

T_M –   Monthly mean of the outdoor temperature in °C

High (T -CH) and Low (T- CL) comfort temperatures are $T_C \pm$ 2°C respectively.

Using above formula and EPW data-base for Coimbatore, the comfort temperature for the summer month of April was found to be 27°C. The comfortable temperature limits were from 25°C to 29°C.

10.1 Development of the control law

Indoor temperature of a house at Coimbatore was simulated using the analytical model proposed earlier. The house was 10 by 10 by 10 feet and thermal properties of the building elements were chosen as given in Table 10.1 below. Specification of the building elements were as recommended by Energy Conservation Building Code for Residential Buildings -2017[97].

Building element	Specification	U [W/m²K]
Wall	1.25cm plaster on both sides,20 cm AAC blocks	0.78
Roof	100mm RCC, 50mm foam concrete, water proofing	1.08

Table 10.1: Thermal properties of building elements used for simulations[97]

Thermal performance of the house was initially simulated without any ventilation. Results are as shown in Figure 10.1 below. It was observed that the indoor temperature was always higher than the comfort high temperature (T_CH).

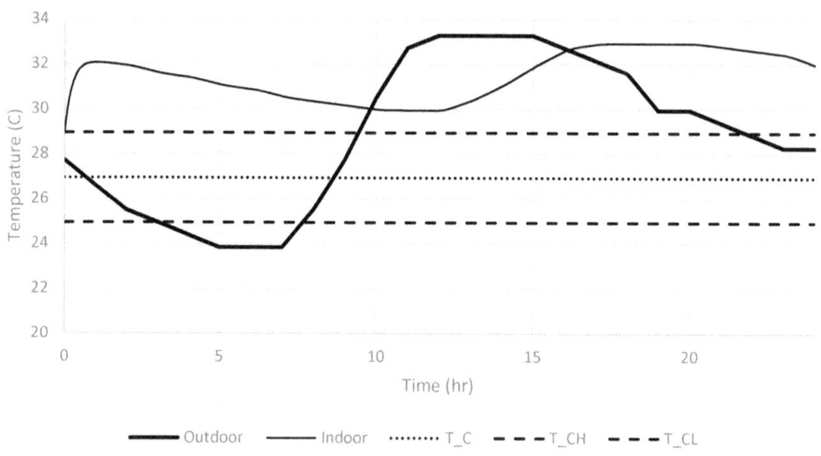

Figure 10.1: Indoor temperature without ventilation

To reduce the indoor temperature, simulation was carried out using only nocturnal cooling, similar to Pablo La Roche scheme[44]. Results are as shown in Figure 10.2 below.

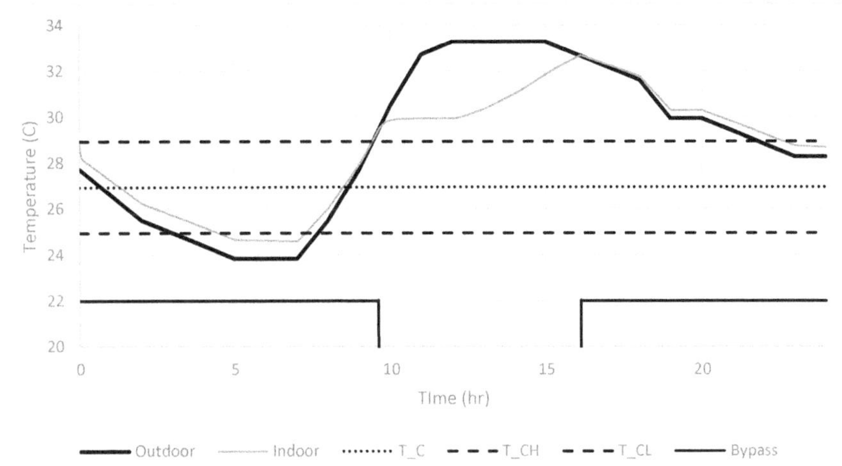

Figure 10.2: Indoor temperature with nocturnal cooling

Nocturnal cooling largely resulted in a reduction in the indoor temperature. But there were periods (10:00h to 22:00h) when the indoor temperature was above the comfort high temperature. A combination of both EAHE and Nocturnal Cooling (Bypass) was then tried and the indoor temperature was observed as shown in Figure 10.3 below.

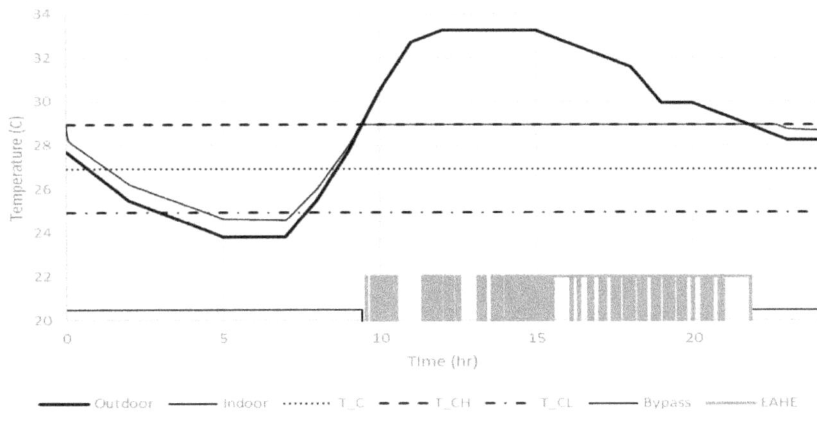

Figure 10.3: Indoor temperature with both nocturnal cooling and EAHE

The indoor temperature was now within comfort limits and also from 1000h to 2400h found to be very close the upper limit of the comfort temperature. Bypass and EAHE 'ON' time of approximately 08 and 16 hours respectively was needed, to maintain the indoor temperature within comfort limits.

Artificial Neural Network (ANN) aided control law:

In order to overcome above mentioned problems, the ANN-heat-transfer model discussed earlier was used for predictive control. A flow-chart for the proposed predictive control law is as shown in Figure 10.4 below.

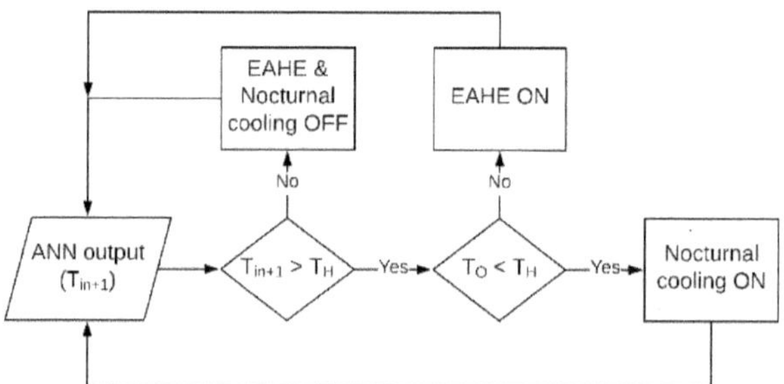

Figure 10.4: ANN based predictive Control Law

Results of the simulation with the control law are as shown in Figure 10.5 below. It was observed that use of ANN (Multi-layer perceptron) aided controller, resulted in substantial reduction in indoor temperature and its variation was found well below the comfort high temperature. The control law activated the bypass and EAHE for 9 and 15 hours respectively.

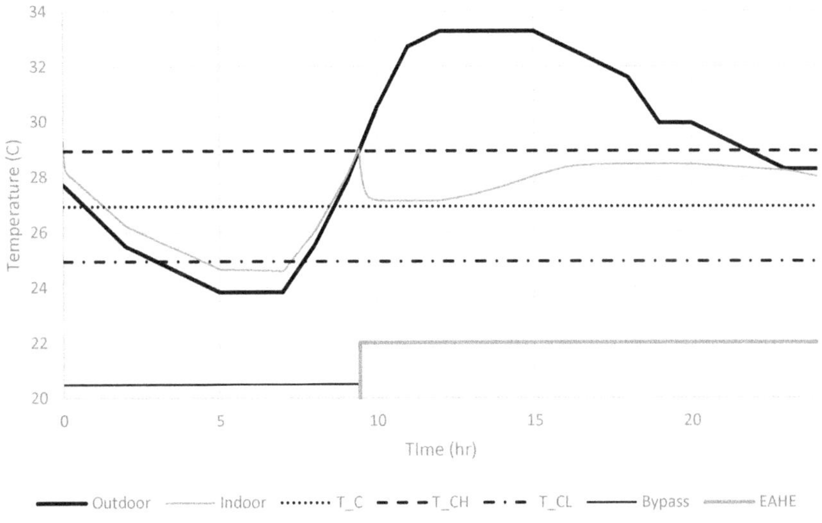

Figure 10.5: Indoor temperature with MLP based ANN (threshold at T_CH)

Simulations were repeated for a Recurrent Neural Network (RNN) based ANN and the results are as shown in Figure 10.6. It was observed that the ON time of EAHE mode reduced to 9 hours compared to 15 hours in the previous simulation (Figure 10.5 above).

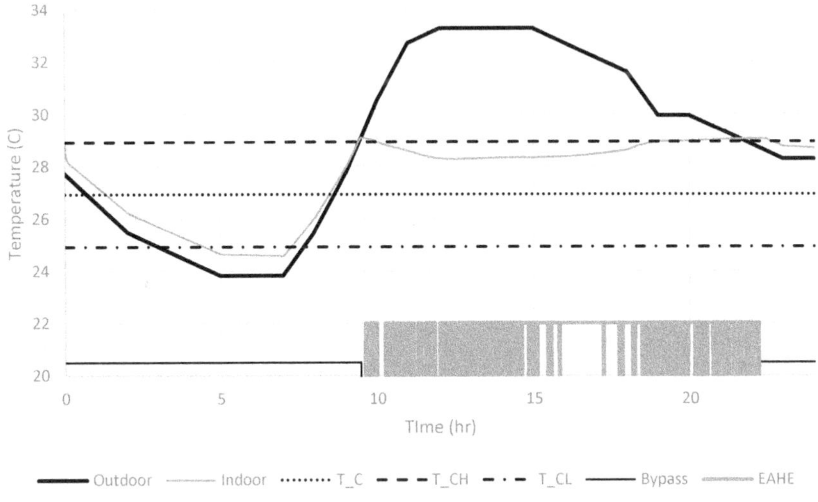

Figure 10.6: Indoor temperature with RNN based ANN (threshold at T_CH)

Simulations with threshold at T-comfort:

Simulations were again carried out by setting the threshold at the comfort temperature (T_C) for both MLP and RNN based ANNs. Results are as shown in Figures 10.7 and 10.8 below. It resulted in reduction average indoor temperature, but EAHE was activated continuously for 16 hours.

Due to change in life styles, houses remain vacant from 0900-1800h, even in developing countries. Hence there is no requirement to keep the EAHE ON for 12-15 hours in a day. Absence of occupants during this period is equivalent to removing 160W of internal load. Indoor temperature variation for such scheme is as shown in Figure 10.9 and 10.10 below. It was observed that MLP was not sensitive to reduction in load, whereas RNN was able to sense variation in internal load and reduced the EAHE 'ON' period to 11.5 hours.

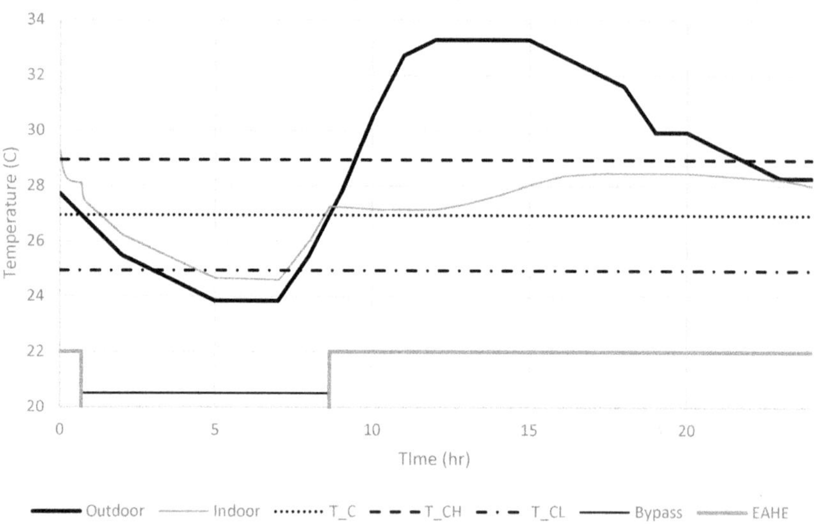

Figure 10.7: Indoor temperature with MLP based ANN (threshold at T_C)

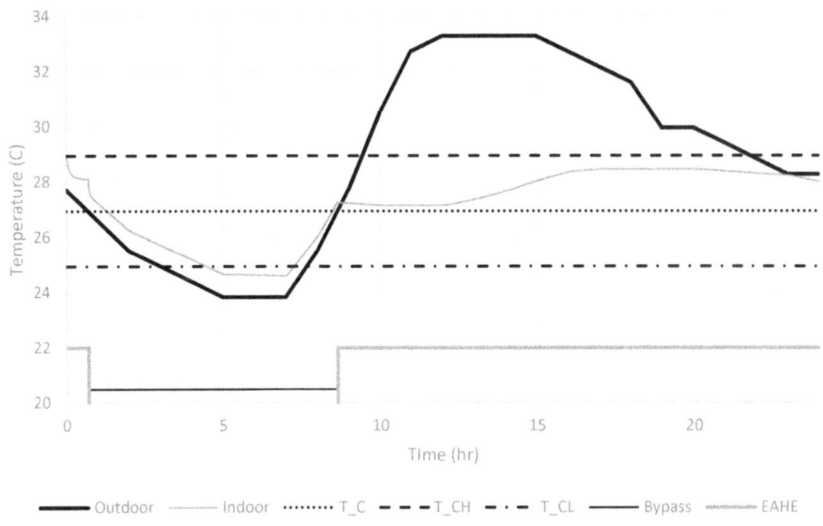

Figure 10.8: Indoor temperature with RNN based ANN (threshold at T_C)

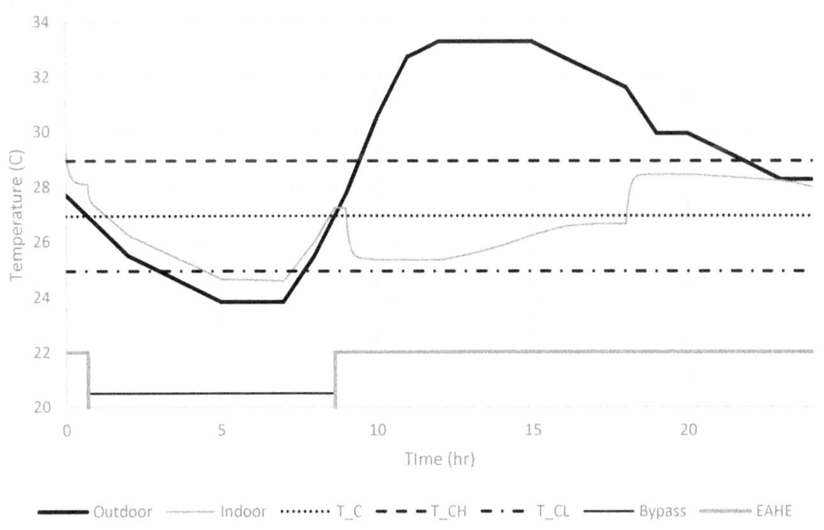

Figure 10.9: Indoor temperature with MLP based ANN (no load from 9:00 to 18:00 with threshold at T_C)

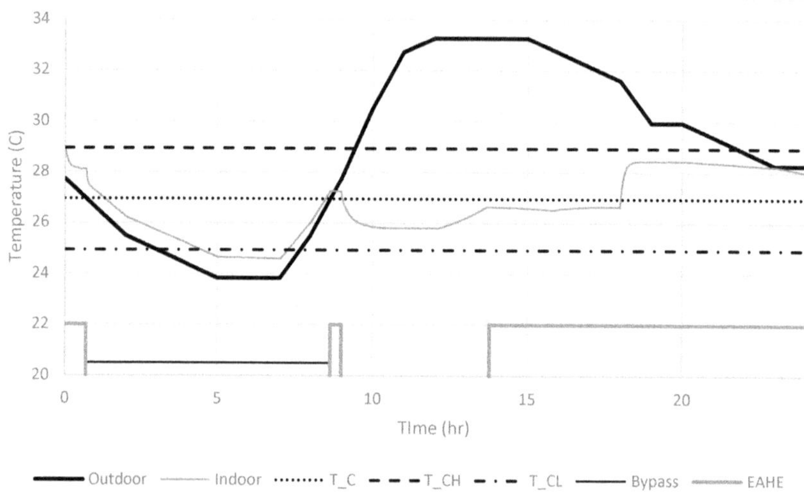

Figure 10.10: Indoor temperature with RNN based ANN (no load from 9:00 to 18:00 with threshold at T_C)

EAHE operation with forced switching OFF of EAHE:

Since the occupancy pattern for a typical family is known, simulations were carried out when EAHE was switched OFF from 1100-1500 h. EAHE was then switched ON three hours before the family returned home. Results are as shown in Figure 10.11. Such a scheme will require only ten and six hours of EAHE operation, for MLP and RNN based ANNs.

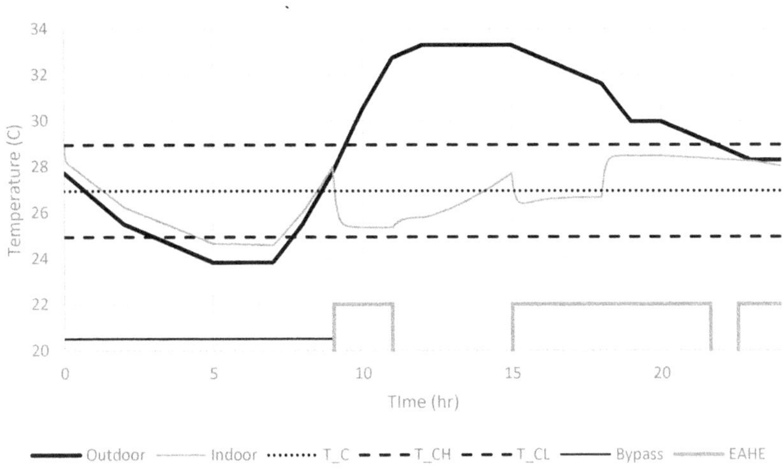

Figure 10.11: MLP with forced switching off EAHE from 11:00 to 15:00h

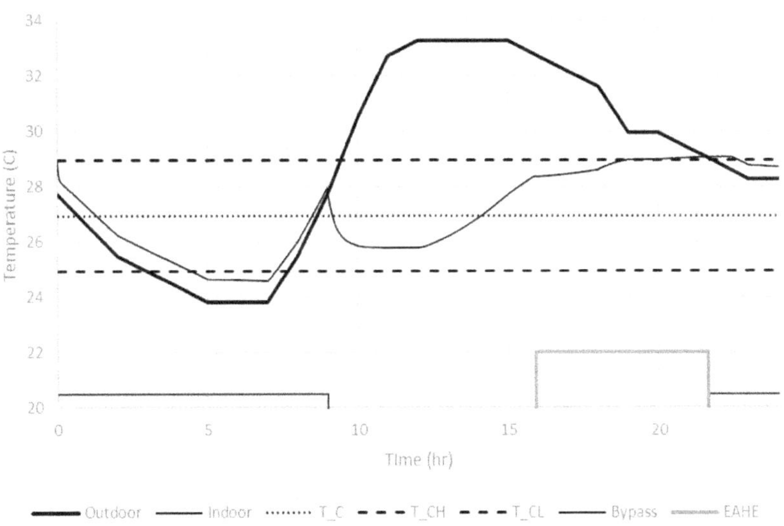

Figure 10.12: RNN with forced switching off EAHE from 11:00 to 15:00h

Performance of various schemes discussed above is summarised in Table 10.2 below. It was observed that RNN based ANN (S9) Model Predictive Control (MPC) law was most energy efficient and maintained comfortable indoor temperature. It was also sensitive to change in occupant load. Forced shut down of EAHE (S10 and S11) was found to be useful if occupancy

pattern is well defined.

No.	Scheme	Activation Logic	EAHE ON (h)	Bypass ON (h)	kWh / Day	Discomfort Hours (h)
S1	No ventilation	______	0	0	0	24
S2	NC only	$T_O < T_{in}$	0	16	0.4	12
S3	NC and EAHE	$T_{in} > T_{CH}$	14	8	1.6	12(close to T_{CH})
S4	NC,EAHE, MPC(MLP)	$T_{in} > T_{CH}$	15	9	1.7	0
S5	NC,EAHE, MPC(RNN)	$T_{in} > T_{CH}$	9	10	1.2	0
S6	NC,EAHE, MPC(MLP)	$T_{in} > T_C$	16	8	1.8	0
S7	NC,EAHE, MPC(RNN)	$T_{in} > T_C$	16	8	1.8	0
S8	NC,EAHE, MPC(MLP) Unoccupied (0900-1800)	$T_{in} > T_C$	16	9	1.8	0
S9	**NC,EAHE, MPC(RNN) Unoccupied (0900-1800)**	$T_{in} > T_C$	**11.5**	**9**	**1.4**	**0**
S10	NC,EAHE, MPC(MLP) Forced shut down (1100-1500h)	$T_{in} > T_C$	10	9	1.2	0
S11	NC,EAHE, MPC(RNN) Forced shut down (1100-1500h)	$T_{in} > T_C$	6	9	0.8	0

Table 10.2: Comparison of various passive cooling schemes

Effect of sudden variations in outdoor temperature:

Surface temperature on the external surfaces of the building envelope can reduce during the day

due to passing clouds or local weather disturbances like thunderstorms. MLP and RNN based

models were subjected to such conditions. Results are as given below in Figures 10.13 and 10.14.

It was found that MLP was insensitive to such changes whereas RNN was able to sense such

conditions and switch OFF the EAHE accordingly.

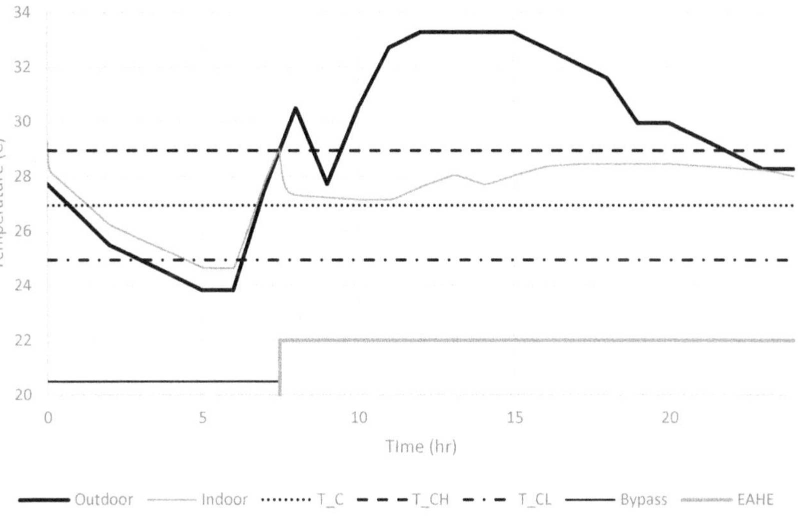

Figure 10.13: MLP with external disturbance (threshold at T-CH)

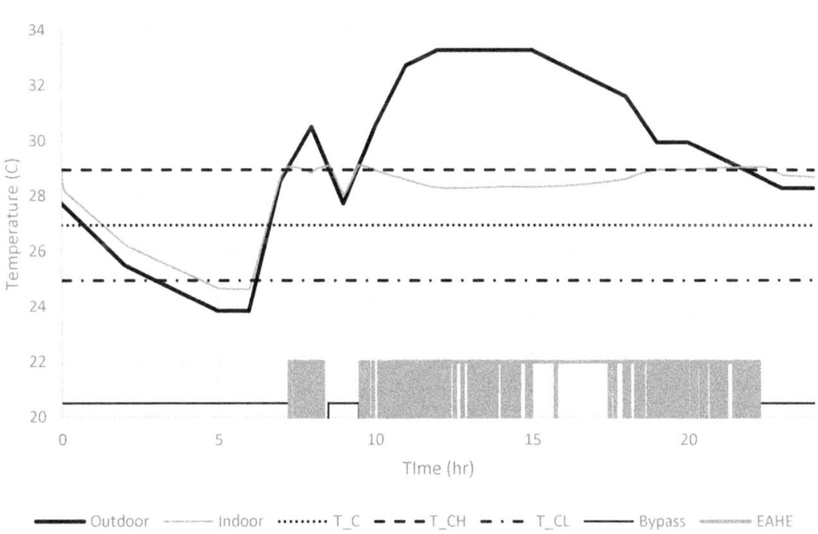

Figure 10.14: RNN with external disturbance (threshold at T-CH)

Simulation of different climate zones:

The climate of Indian Subcontinent can be broadly classified into six hot Koppen zones[98]. In order to ascertain the effectiveness of above mentioned RNN based control law (S9) for different climatic zones, simulations were carried out for cities listed in Table 10.3 below.

City	Koppen climate zone classification	Climate characteristics
Ahmedabad	Bsh	Arid-steppe-hot
Coimbatore	Cwb	Temperate - dry winter - warm summer
Chennai	Aw	Tropical savana
Jaisalmer	Bwh	Arid-desert-hot
Mumbai	Am	Tropical monsoon
Nagpur	Cwa	Temperate - dry winter - hot summer

Table 10.3: Important climate zones of India

Simulations for above-mentioned cities were carried out using climate data available in EPW database and no discomfort periods were observed. Energy consumption and operational cost to maintain indoor comfort temperatures for the hottest month was compared between the proposed cooling system and commercially available three stars rated 1Ton air conditioners (with a constant set point temperature of 26°C for all locations). Results are as shown in Table 10.4 below.

City	Energy (kWh)/Day		Operational Cost/Day		% difference
	Proposed system	AC (1Ton)	Proposed system	AC (1Ton)	
Ahmedabad	5.4	16.2	32.4	97.2	67
Coimbatore	1.4	10.0	8.4	60.0	86
Chennai	3.9	13.0	23.4	77.7	70
Jaisalmer	4.7	16.7	28.2	100.4	72
Mumbai	5.7	18.4	34.2	110.2	69
Nagpur	6.3	18.9	37.8	113.4	67

Table 10.4: Comparison of energy consumption and operational cost

System Cost: The cost of the proposed system is only Rs 10,000/- as against a conventional 3 star air conditioner which costs Rs 25,000/-. Cost of major items is given in Table 10.5.

Item	Item Cost
Centrifugal blower (0.5 hp)	Rs 3000
PVC piping	Rs 1000
Gravel	Rs 2500
Controller Board (Arduino)	Rs 500
Sensors (two temperature sensors , one RTC)	Rs 1000
Labour for trench	Rs 2000

Table 10.5: Item wise Cost for the proposed system

Chapter 11

Results and Conclusions

Thermal comfort was a basic physiological need identified by Maslow[77]. In the developed world, this need has been satisfied by air-conditioned buildings, which followed the same standard, irrespective of the geographical location of the building. Such practices have resulted in high energy consumption, which in turn has contributed to greenhouse effects. Passive cooling techniques have not found acceptance in the society and conventional air-conditioners were sought after to provide thermal comfort

An attempt has been made to identify reasons for lack of popularity of passive cooling technique, improve and integrate existing techniques. Based on studies carried out, an easily implementable, low-cost alternative to conventional air-conditioner is proposed. Important contributions of the thesis are as follows:

11.1 Thermal comfort model

It was found that the root cause for high energy consumption towards cooling was due to setting of the air-conditioner thermostat at a constant indoor temperature, irrespective of the geographical location.

Subsequently ASHRAE[16] introduced the adaptive comfort model, as a standard for naturally ventilated buildings. It was found that this standard can only be used under laboratory conditions, where simultaneous measurement of all internal surface temperatures can be carried out.

With the help of experiments, it was found that such measurements were not required in passively cooled buildings. After studying different thermal comfort models, a suitable adaptive thermal comfort model for passively cooled buildings, which required measurement of only ambient outdoor temperature, was identified.

11.2 Improvements in Passive and natural cooling methods:

11.2.1 Optimization of building envelope

The most important aspect of passive cooling is to reduce heat gain in to the building. After studying buildings plans with different ratios of length (L) to width (W), it was found out that an L/W ratio close to golden ratio (1.618) resulted in least heat gain. Further increased time lag of the building material, was identified as an important property to reduce heat gain.

11.2.2 Earth Air Heat Exchanger (EAHE)

Although the physics of EAHE was well documented, it has rarely been used for small households. High installation cost towards digging of a 4m deep trench was identified as the main reason for lack of popularity of this technique. With the help of analytical study, simulations and experiment, a novel method of achieving satisfactory EAHE cooling, at a lesser cost has been demonstrated.

11.2.3 Improvement of natural ventilation with wing walls

An indoor temperature of 35°C can be tolerated, if the indoor air velocity is at least 1 m/s. At present, buildings generally do not permit such indoor air velocities.

Wing walls which are projections on the external surface near the windows improved the indoor air velocities. Literature available was only for wing walls for windows placed on the same wall.

Wing walls placed on adjacent walls were studied using CFD. Parametric studies on wing wall height and width was carried out. An inverted 'L' configuration was proposed. It improved the window inlet velocity to 3.7 m/s compared to 2.64 m/s for conventional windows.

11.2.4 Study on night time ventilation (nocturnal cooling)

Natural ventilation can reduce indoor temperature, only if the outdoor temperature is less than that indoor temperature. Based on measurements, such conditions were observed generally between 1800h in the evenings to 0800h in the mornings next day and indicated that natural ventilation was effective only during night time (nocturnal cooling). Because of reduction in wind velocity during both summers and nights and also due to barriers for natural ventilation, it was concluded that natural ventilation alone was not a viable passive cooling technique.

Controlled nocturnal cooling: was studied by researchers to improve natural ventilation and to mitigate some of its barriers. Though this technique reduced energy consumption compared to air conditioned buildings, it has not become popular because it did not prevent discomfort hours (on a 24h basis), during summers in most of the climatic zones. Hence it was felt that both controlled nocturnal cooling and EAHE should be used to ensure comfortable indoor temperature throughout the year, in all climatic zones.

11.3 Integration of different passive cooling techniques

11.3.1 Formulation of analytical and ANN based model

A gap has been identified between modelling, simulation and utilisation by the construction industry. A new building heat transfer model was formulated which required only easily measurable inputs and was suitable for naturally ventilated buildings. This model was further improved with two versions of Artificial Neural Network (ANN) based models.

11.3.2 Development of control law

With the help of the proposed heat transfer model, a control law was formulated to integrate different passive cooling techniques. The control law mimicked the most important aspect of human nature i.e. learning over a period of time. Advantages of Recurrent Neural Network (RNN) based ANN model over that of Multilayer Perceptron (MLP) based ANN were identified for controlling the indoor temperature. The control law was found to be effective for all hot climatic zones of India. Hence it can be generalized for all tropical countries.

The proposed system has met all the objectives (Chapter 3) and has resulted in a low cost cooling system compared to conventional air-conditioners. The installation and operating cost of the proposed system were 60% and 70% lower compared to that of a conventional air-conditioner.

11.3.3 Limitation and scope for future work:

- The study was restricted to single room houses in rural and semi urban locations, where sufficient space is available for installing an EAHE.

- Installation of EAHE is not possible in apartments located in urban areas. Hence reduction of heat gain is the only option. This will require development of better insulation materials

- This study can be extended to multi-zone models.

- PVC pipes were used for experiments on EAHE. Use of Aluminium pipes which have better thermal conductivity can be explored to improve the performance of the EAHE.

- Use of constructed wet lands can be explored for soil conditioning.